SpringerBriefs in Biochemistry and Molecular Biology

For further volumes:
http://www.springer.com/series/10196

Anirban Banerji

Fractal Symmetry
of Protein Interior

 Springer

Anirban Banerji
Bioinformatics Centre
University of Pune
Pune
India

ISSN 2211-9353   ISSN 2211-9361   (electronic)
ISBN 978-3-0348-0650-3   ISBN 978-3-0348-0651-0   (eBook)
DOI 10.1007/978-3-0348-0651-0
Springer Basel Heidelberg New York Dordrecht London

Library of Congress Control Number: 2013934381

Printed on acid-free paper

Springer is part of Springer Science+Business Media (www.springer.com)

This book is dedicated to my parents,
Leena Banerji and Swapan Banerji

Preface

The essential question that fractal dimensions attempt to answer is about the *scales* in Nature. This is not exactly a new question. From Euclid to Poincare to Mandelbrot, many philosophers, mathematicians, physicists had thought about transformation, dilation, and contraction of scales. Why is it important to investigate this problem? It is important not merely to document the characteristic scales observed in various phenomena of real world, there's something more fundamental than that. Our understanding of physics of a system might at times be dependent on our selection of scales (or resolutions) at which we investigate a system. Hence an idea about *scaling* and (possible) *scale-invariance* assumes significance when we attempt to study any physical system. For a system as non-idealistic and complex as a protein is, the question of studying *scale-invariance*, becomes particularly important. One may (realistically) hope that investigations related to *scaling* of biophysical and biochemical behaviors may one day help us to formulate a fundamental theory about protein biophysics; which, in turn, may help us to understand proteins from a set of fundamental principles.

In particular, this (small) book will attempt to investigate the diverse facets of the question: *among the various scales at which we describe protein biophysical and biochemical phenomena, which scales are scale-invariant?* A detailed exploration of this problem will perhaps help us to investigate an even deeper question: *how does evolution determine the fundamental scales of biological world?*

There are quite a few fantastic textbooks on mathematics of fractals and renormalization theory. These topics will not be dealt with in the present work. Furthermore, this book will not talk about the various methodologies and results associated with multifractal description of protein properties. Instead of being mathematically pedagogic, the approach adopted throughout the course of this book will be qualitative, with clear onus on discussions related to protein structures. Having said that, since it is absolutely obligatory for the readers to understand the concept of fractals, a (largely) qualitative and (somewhat) long discussion about fractal dimensions is kept in the introduction of the book. Furthermore, since it is imperative to have slight idea of polymer and polymer collapse for anybody studying Fractal Dimension (FD)-based treatments of proteins, a brief section to compare and contrast polymer collapse and protein folding has been provided.

This book attempts to collate and classify the various FD-based approaches that have been employed over the years to study protein interior biophysical properties into distinct clusters. It then presents an account of cases where FD-based methodologies have successfully contributed to protein interior research. In this context, a thorough documentation of accurate predictions made from the spectrum of FD-based studies is provided. However, fractals are no panacea, and they cannot suggest magic solutions to all the problems of protein biophysics. Consequently, a third aim of this book is to examine the intrinsic limitations of FD-based measures. Finally, with a balanced assessment of the entire framework, the book attempts to identify some of the outstanding questions for which the application of FD-based investigations may help in deciphering deep and unexpected facets of protein interior organization.

This book would not have come to being without the direct or indirect support from countably infinite number of well-wishers of mine. I thank them all. I would like to take this opportunity to thank all those who asked me questions about fractals and proteins over the years, and especially to those whom I could not answer for my lack of knowledge and/or conception. These questions stayed with me and forced me to spend more time in library and in my work. In a way, it is through these interactions that I learned whatever little I have learned about fractals and proteins. Thank you all.

This preface will be incomplete if I do not mention the roles of the editors, Dr. Beatrice Menz and (later on) Dr. Jutta Lindenborn during the course of writing the book. They were always supportive, always encouraging. I thank Dr. Menz especially for entrusting me with the responsibility of writing this book.

There is nothing more helpful than to listen to constructive criticisms; thus, if you feel like criticizing and/or discussing and/or (in the unlikely case) appreciating any aspect of whatever that is written in the book, please drop me an email ([anirbanab][what goes here][gmail.com]). I will be delighted to talk to you and learn from you.

Pune, India   Anirban Banerji

Contents

1 Introduction... 1
  1.1 How did it All Begin? (Brief History of Fractal Dimension)........ 1
  1.2 Introduction to Fractals 10
  1.3 Misconceptions About Fractals 13
  References.. 16

2 Studying Protein Interior with Fractal Dimension 19
  2.1 Why, at All, Does One Need Fractal Dimension to Study
    Protein Interior?... 19
  2.2 Schools of Protein Interior Fractal Studies..................... 21
    2.2.1 Approaches with Backbone Connectivity 22
    2.2.2 Approach with Residue (and Atom) Distribution.......... 31
    2.2.3 Approach with Correlation Dimension Analysis 33
    2.2.4 Approaches with RGT 40
    2.2.5 Approach with the Spectral Dimension 42
  2.3 Results Obtained with Fractal Dimension-Based Investigations 52
    2.3.1 FD-Based Protein Conformation Studies................ 52
    2.3.2 FD-Based Ion-Channel Kinetics Studies 53
    2.3.3 FD-Based Attempts to Relate Protein Structure
       and Dynamics...................................... 53
    2.3.4 Studies in Fractal Kinetics 54
    2.3.5 FD-Based Results on Protein Structure 55
  2.4 Gaining New Knowledge About Protein Interior with FD-Based
    Investigations 58
    2.4.1 Implementation of CD-Based Constructs................ 58
    2.4.2 CD-Based Investigation of Local and Global Dependencies
       Amongst Peptide Dipoles Units...................... 61
    2.4.3 CD Amongst Charged Amino Acids Across All Seven
       SCOP Classes..................................... 64
    2.4.4 CD Amongst Aromatic Amino Acids Across All Seven
       SCOP Classes..................................... 65
    2.4.5 CD Amongst Hydrophobic Amino Acids Across All Seven
       SCOP Classes..................................... 67

2.4.6   General Inferences About Protein Stability from SCOP
        Class Wide Separate Analyses of CD Between Charged
        Residues, Peptide Dipole, Aromatic Amino Acids,
        and Hydrophobic Amino Acids   68
2.4.7   Correlation Dimension-Based Investigation of
        Dependency Distribution Amongst Hydrophobic
        Residues, Charged Residues, and Residues with
        π-Electron Clouds   69
2.5   Problems with FD, Necessary Precautions While Working
      with FD..   70
2.6   New Directions in Fractal Dimension-Based Protein Interior
      Research.......................................   71
2.7   Conclusion   73
References...   73

Chapter 1
Introduction

Abstract Here we introduce ourselves to fractal dimensions. The idea of fractional dimension had a background that dated back to more than a century before Mandelbrot's works. Concepts, when viewed in their historical contexts, often present themselves as easier ideas to understand. I will, therefore, (very) briefly, talk about the history of fractal dimensions. But I am not a historian and I make no claim that every piece of historical fact is covered here; nor do I attempt to cover all the aspects of evolution of ideas that have finally given birth to concept of fractal dimension. Although these ideas are entirely mathematical (involving not a morsel of Physics or Biology), mathematical expositions have been carefully avoided in this book, unless the use of some expressions becomes absolutely obligatory. To help the interested readers, numerous references are suggested. Various misconceptions often plague students and researchers in the field of fractals;—a small attempt, therefore, is made to point out some of the most notable misconceptions.

1.1 How did it All Begin? (Brief History of Fractal Dimension)

Shortly after the World War I, in 1919 (1918, according to some others) Martin Jasek, a Czech secondary school teacher, discovered an old manuscript in National Library of Vienna. Some 2 years down the line, in the winter of 1921–1922, Jasek chose to talk about the content of this worn out manuscript, which dated back to 1830. Lectures were read on December 3, 1921, January 14 and December 2, 1922. The 1921–1922 Martin Jasek lecture series (Jasek 1922) shook the then-contemporary mathematician community, because the discovered manuscript revealed a breathtakingly courageous stream of thoughts of a Czech theologist named Bernard Bolzano. Bolzano presented the first example of a function that is *continuous at every point but differentiable nowhere*, and he did so way back in 1830! The "pathological cases" (that's how mathematicians used to refer to these functions) were not entirely unusual findings in 1921, however, conceiving one such case in 1830—was shockingly amazing to know (Folta 1966; Nový 1981).

But what is *so* remarkable in thinking about curves that are continuous at every point but differentiable nowhere? To the extent that studies of such systems came under the crowning label of "pathological cases"? Well, naively, one expects that a continuous function must have a derivative; or else, the set of points where it is not differentiable will be very few. For example, think of the function $f(x) = |x|$. The plot of absolute value function shows that $f(x)$ is equal to x when x is positive and $f(x)$ is equal to $-x$ when x is negative. At the point $x = 0$, the function has a kink, like the letter 'V' has. At that particular point ($x = 0$), viz. at the kink, the absolute value function is nondifferentiable. But compared to the total number of possible points that make up the graph for this function, this is just one particular point. That, however, was not the case with Bolzano's function; it was continuous at every point but differentiable nowhere. The fact that such a function was conceived at a time when everyone of the great mathematical minds of that time was absolutely convinced of the infallibility of the 'continuity implies differentiability' paradigm, makes Bolzano's effort truly remarkable. Interestingly, Bolzano's work was not published then.

But Bolzano was not alone. Some 30 years down the line, sometime around 1860, Charles Cellérier, a professor at the University of Geneva, Switzerland, independently formulated another *continuous-but-nowhere-differentiable* function. But, in striking similarity to the case of Bolzano, the work was not published then. Cellérier's function was published posthumously in 1890 (Cellérier1890).

There are traceable signs, however, that times were changing slowly. In a classroom lectures in 1861 (at the very latest), one of the doyens of nineteenth century mathematics, Karl Weierstrass, talked about functions that can be continuous but nowhere differentiable (Boyer and Merzbach 1989; MacTutor 2000). Later on, in a 1972 paper (to which we will come back shortly), Weierstrass stated categorically that it was Bernhard Riemann (one of the greatest mathematicians), and not he himself, who was the first to definitely assert that, the infinite series: $\sum_{n=1}^{\infty} \frac{\sin(n^2 x)}{n^2}$, which is manifestly continuous, is not differentiable. Most importantly, Riemann did it in 1861, or sometime before than that. Unfortunately, it is not exactly evident that Riemann asserted or proved the fact that the series is indeed continuous but nowhere differentiable. (Interested readers are referred to (Neuenschwander 1978) and (Butzer and Stark 1986), for captivating views on this topic).

On page 560 of the 1872 'Monthly Reports of the Royal Prussian Academy of Science in Berlin', one finds a mention that on July 18, "Hr. Weierstrass las über stetige Funktionen ohne bestimmte Differentialquotienten" (that is, Mr. Weierstrass has read [a paper] about continuous functions without definite [i.e., well-defined] derivatives [to members of the Academy]) (see Weierstrass function, hypertext file http://en.wikipedia.org/wiki/Weierstrass_function). It is significant to note that this work was not published in the Monatsberichte. In the second volume of Weierstrass' Mathematische Werke (published in 1895, 23 long years after his Berlin Academy talk) one finds a paper that is generally believed to be the record of Weierstrass' talk given to the Berlin Academy of Sciences on July 18, 1872. It is not known when exactly this paper was first formally written. Anyway, in this landmark work, Weierstrass proved that the function: $f(x) = \sum_{n=0}^{\infty} b^n \cos(a^n x\pi)$ is continuous, but it is nowhere differentiable; if $b \in (0,1)$, a is an odd integer, and $ab > 1 + \left(\frac{3\pi}{2}\right)$. The 1872 Weierstrass talk is widely considered as the first publicly available record of thoughts on 'continuous but nowhere differentiable' function.

But world of mathematics was not sitting idle between 1872 and 1895. In that same year, 1872, another German mathematician Felix Klein, suggested viewing geometry as "the study of the properties of a space which are invariant under a given group of transformations" (Mumford et al. 2002). He went on to suggest that geometry needed to include not only idealistic shapes (that is, spheres and cubes etc.) but also the irregular shapes and even, movements (for example, movements depicted in Brownian motion). Klein worked with ideas of similar/symmetric objects and transformations/movements in the plane and realized that studying the features of the object that were left unchanged by the transformations—is an extremely significant idea. He described symmetry as a balance created by similar repetitions. In an exhibition of startlingly deep insight, he realized that iteration of the same motion over and over, may account for creation of a pattern or an object that is symmetrical with respect to motion;—that is, an object where the individual points of the figure change position but the pattern or the shape of the object, as a whole, remains unchanged. In these, prophetic realizations lay the ground for modern-day recursion, a pivot to construct deterministic fractals. Although we won't talk about these aspects of fractals in the present book, interested readers will be benefitted by referring to (Mumford et al. 2002) and (Gleick 1987).

Two other important developments took place between 1872 and 1895 too. In the first of them, in 1883, Georg Cantor (another German mathematician, renowned for his fundamental contributions to set theory), while elaborating on a perfect set that is nowhere dense, fleetingly mentioned about the ternary set (Cantor 1883). In ternary set one takes a unit interval $[0,1]$, and removes the middle-third portion; and then taking each of the two remaining parts, removes the middle-third from each one of them, and carries on performing the same operation, recursively. The set, obtained finally, is a fractal,—now known universally as the 'Cantor set'. Although many others had discovered such a set independently (see Cantor set, hypertext file http://en.wikipedia.org/wiki/Cantor_ternary_set), we will refrain ourselves from talking about them. The 'Cantor set', is a typical "pathological case". To understand the reasons for embarrassment of the then-contemporary mathematicians, let us sneak a look at some of the "paradoxical qualities" of Cantor set. Say, one zooms in on the parts that have been removed. The length that has been removed at each section is given by: $\frac{2^n}{3^{n+1}}$, where 'n' denotes the number of iterations. To obtain the global idea, let us sum this fraction between zero to infinity, viz. $\sum_{n=0}^{\infty} \frac{2^n}{3^{n+1}}$. Upon performing this summation with standard formula, one obtains $\sum_{n=0}^{\infty} \frac{2^n}{3^{n+1}} = \frac{1}{3}\left(\frac{1}{1-\frac{2}{3}}\right) = 1$! Which suggests that even after removing an infinite number of middle thirds, the length of the line segments, remains the same as it was in the very beginning!—Nobody had ever seen anything more bizarre and strange in 2,000 years of geometry studies (Gleick 1987).

Then there was the question of dimensionality of this object. Today we know that the dimensionality of the cantor dusts is 0.6309; but during those days, people had torrid time in trying (and failing) to classify the Cantor set as either an object with 0 dimension or an object with 1 dimension;—when, in fact, it is neither. Quite understandably, such a case was labeled "pathological".

But to the agony of those who considered Cantor set to be "pathological", motivated by Cantor's set, an Italian mathematician Giuseppe Peano discovered the first space-filling curve in 1890 (Peano 1890). The word 'space-filling' implies just how densely a particular space is being filled by a curve. For example, have a look at Figs. 1.1, 1.2, 1.3, 1.4, 1.5, and 1.6. Although curves in Figs. 1.1–1.6—all attempt to traverse between one fixed point to another fixed point in a fixed "space", curve in Fig. 1.1 can be found out as the least space-filling (that is, least dense among

Fig. 1.1 Demonstration of space-filling properties of a *curve*

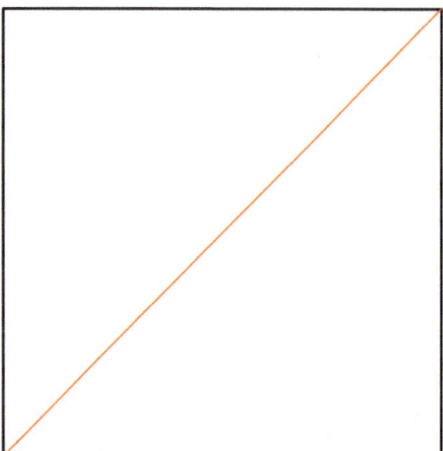

Fig. 1.2 Demonstration of space-filling properties of a *curve*

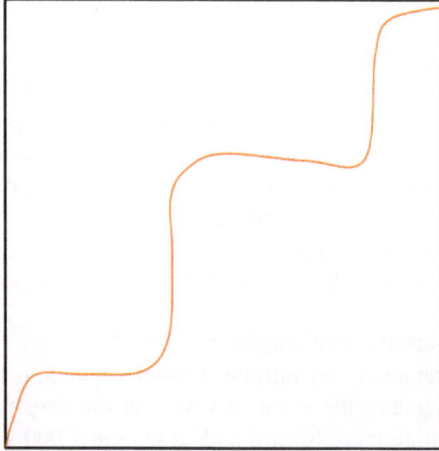

Fig. 1.3 Demonstration of space-filling properties of a *curve*

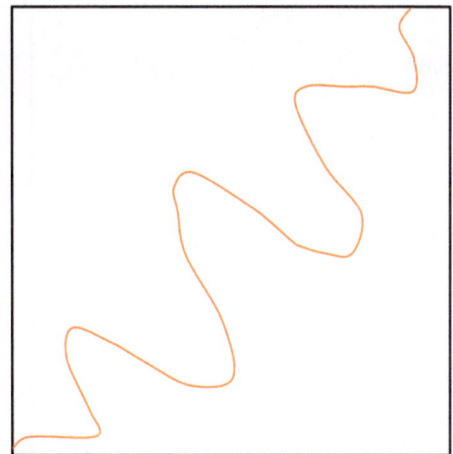

Fig. 1.4 Demonstration of space-filling properties of a *curve*

all the options), whereas Fig. 1.6 can easily be identified as the most space-filling (most dense among all the options). Coming back to the original discussion, while with Cantor's set, one arrives at an object with dimensionality in between that of a point and a straight line, with Peano's curve one arrives at an area starting with a line. Such a curve (a 1-dimensional object) that fills up an area (that is, 2 dimensional space) by touching every point in the space, was immediately labeled as another "pathological case". The very next year, in 1891, David Hilbert (one of the most influential mathematicians of all times) discovered another space-filling curve (Hilbert 1891). Interestingly, both of these curves happened to be continuous but nowhere differentiable. (However, one should not conclude from these examples that all the space-filling curves belong to 'nowhere differentiable' category; Lebesgue's space-filing curve, for example, is differentiable almost everywhere).

Fig. 1.5 Demonstration of
space-filling properties of a
curve

Fig. 1.6 Demonstration of
space-filling properties of a
curve

   To present another example of a pathological case, we will talk briefly about
the Koch curve. A Koch curve is generated by a replacement rule, the rule being:
at each step, replace the middle (1/3)rd of each line segment with two sides of a triangle, where the triangle has sides of length equal to the replaced segment. Diagram
(Fig. 1.7) provided herewith shows the first few steps how to generate a Koch Curve.
We find that, if the length of the initial line segment is l, the length L_{Koch} of the Koch
curve at the nth step will be given by: $L_{Koch} = (4/3)^n.l$. Since L_{Koch} keeps on increas-
ing, the Koch curve, eventually acquires an infinite length; though bounded in a finite
region of space (that is, in the present case, a piece of page of this book).
   The dimension of an object is a measure of its complexity. Intuitively, in
Euclidean space, the dimension of a set is the amount of information, necessary, and sufficient to identify any point in the set. Loosely, and I repeat, loosely,
one can think of this idea of dimensionality as the same obtained with topological

Fig. 1.7 Steps to generate the Koch *curve*

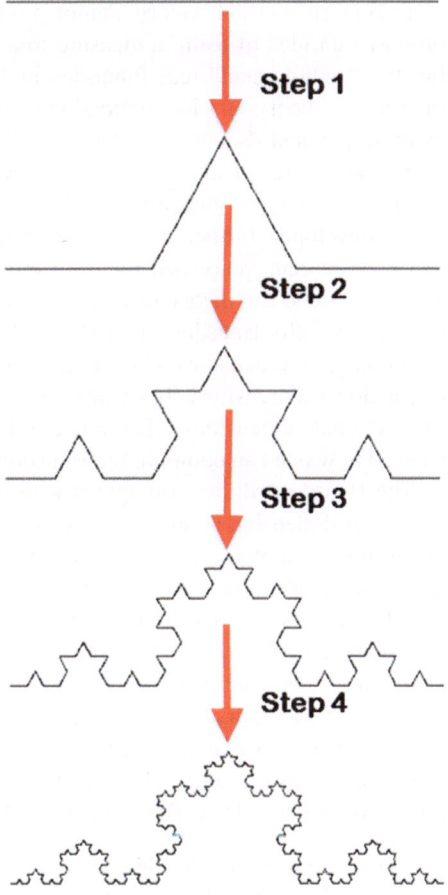

dimensions. The topological dimension of a set is always an integer, and is 0 if the set is totally disconnected. Or alternatively posed, traditionally, points are considered to be objects with topological dimension 0 as they have no size at all (no length, no width, and no height); which means, without the fact that they are existing somewhere in space, there is nothing to specify their position. Curves have topological dimension 1 because they can be parameterized by a single real coordinate, and (smooth well-behaved) surfaces have topological dimension 2 since two coordinates describe every point on such a surface. The topological dimension of $\mathbb{R}^n$ is n. (This can be proven, but we will not venture to do so here. To learn about various facets of topological dimensions and relations between topological dimension and fractal dimension, interested readers can look at (Edgar 1990),—a classic). However, this intuitive concept of dimension is not completely adequate. For example, how can one assign dimensionality to Peano curve, which is a curve but it touches every point in the 2 dimensional object unit square?—Thus, although intuitive, topological dimensions are not necessarily the most adequate constructs.

During 1874–1884, Georg Cantor produced a series of remarkable papers and proposed an idea of using a measure to extend the notion of length. While studying the "pathological" real functions in 1895, Emile Borel (one of the pioneers of measure theory and its application to probability theory) found this idea to be extremely potent. Some years later, in 1902, these ideas were extended by Henri Lebesgue (a French mathematician, known for his fundamental contributions to the theory of integration) in his work on Lebesgue integral. The ideas of Lebesgue where developed further by Constantin Caratheodory (a German mathematician of Greek descent, renowned for contributions to various fields of mathematics) in 1914, to extend the theory to lengths in arbitrary spaces. This idea was later generalized by Felix Hausdorff in 1919 in (Hausdorff 1919), who extended it further to nonintegral dimensions. Hausdorff's contribution is the foundation of the theory of fractional dimensions. Ten years afterward, in a paper 'On linear sets of points of fractional dimensions' (Besicovitch 1929), Abram Samoilovitch Besicovitch showed how one can compute the Hausdorff dimension for highly irregular sets.

The Hausdorff dimension agrees with our intuition for simple sets. Points have Hausdorff dimension 0, simple curves have dimension 1, and simple surfaces have dimension 2. But being an extremely robust construct, it can quantify the dimensionality of complex curves too, which may have dimension larger than 1. For example, the Peano curve has Hausdorff dimension 2. Similarly, for complex surfaces Hausdorff dimension can be larger than 2. Most importantly, the *Hausdorff dimension of a set need not be an integer, it can be an arbitrary real number.* This inclusive framework, characterized especially by the last sentence, changed the way in which geometry was traditionally viewed. Further progress in this line of thought was made when a framework to relate the fractional dimensions to Geometry was proposed sometime later (Besicovitch and Ursell 1937); (Marstrand 1954).

> Why is geometry often described as "cold" and "dry"? One reason lies in its inability to describe the shape of a cloud, a mountain, or a tree. Clouds are not spheres, mountains not cones, coastlines are not circles, and bark is not smooth, nor does lightning travel in a straight line.

... Said the opening of 'The Fractal Geometry of Nature', a 1982 book written by Mandelbrot (1982). These were courageous and prophetic doubts (and statements). But Mandelbrot was thinking in these lines for a quite a while. In his 1967 paper (Mandelbrot 1967) he talked about similarity dimension. Here, Mandelbrot formalized the work of Lewis Fry Richardson, who noticed (Richardson 1961) that the length of a coastline depends on the unit of measurement used. Mandelbrot realized that the theory of fractional dimensions and self-similarity could be used not only in mathematics, but in various other branches of science as well. Thinking along this line, in one sharp piece of realization, Mandelbrot recognized that geometrically self-similar figures will seldom be found in nature, but nature will embody an inexhaustible number of cases with statistical forms of self-similarity. The ubiquity of statistical self-similarity and the reasons why fractal geometry will be more appropriate and honest in describing the nature than classical Euclidean geometry has been,—is elaborated in details in his 1982 book

(Mandelbrot 1982). Interested readers, while searching for discussions on topological dimension, geometric self-similarity, statistical self-similarity—may also benefit from reading (Kaye 1994). Figures 1.8 and 1.9 show the difference between a case of geometric self-similarity and statistical self-similarity. Figure 1.8 shows an example of exact (geometrical) self-similarity, displayed vividly by the 'Koch-curve', whom we've met before. Figure 1.9, in contrast, is an example of statistical self-similarity, where the roughness profile of the Earth's surface is shown. Both diagrams have been reproduced here courtesy of Professor Paul Bourke (Bourke, hypertext file, last accessed on Sept., 2012).

Now then, how to formally define a fractal?—I'm afraid, this question cannot be answered in the present book; because a discussion on formal definition of fractals require a thorough foundation on dimensional analysis amongst many other mathematical pre-requisites. A somewhat-rigorous and possibly formal way of answering

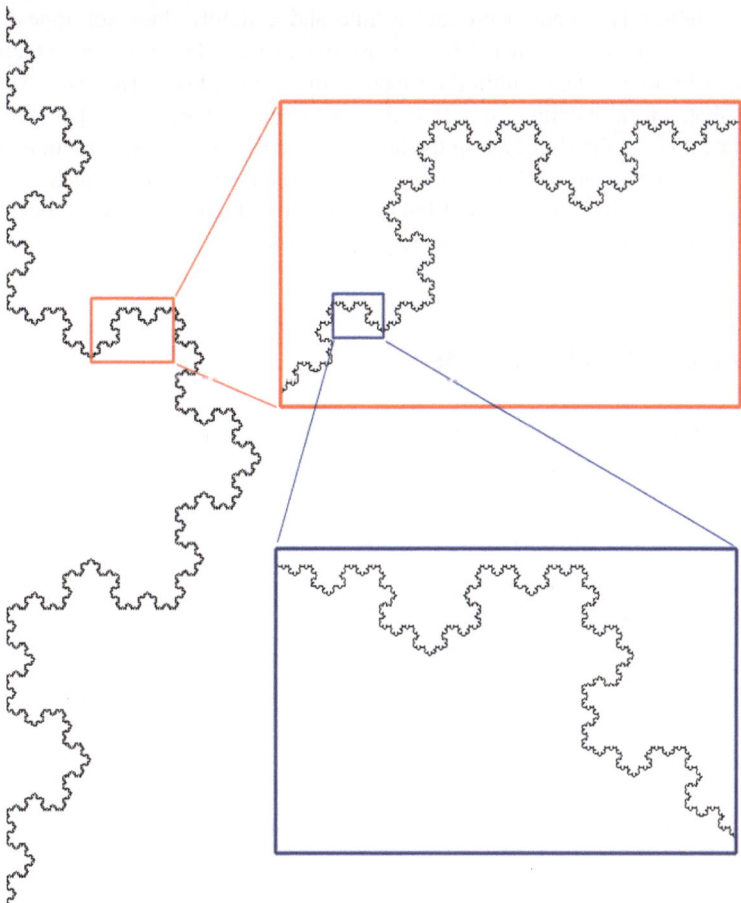

Fig. 1.8 Demonstration of geometric self-similarity

Fig. 1.9 Demonstration of
statistical self-similarity

the question 'what is a fractal?' can be: it is a fractal that represents the attractor
of an iterated function system on some compact subset of a complete metric space.
Unfortunately, this sounds a bit cataclysmic and certainly does not appear to be
friendly to anyone not familiar with dimensional analysis. Discussions on these topics cannot be undertaken within the scope of the present book. However, the good
news is that, a formal definition of fractals is not really necessary to either learn or to
appreciate their use in deciphering the latent symmetries in protein structure organization. What, however, is of utmost necessity for us, is to develop a qualitative, and
yet strong conceptual ground about fractals, which will teach us how to identify the
possible fractal nature in an object merely by observing its properties.

1.2 Introduction to Fractals

In his 1975 paper (Mandelbrot 1975) Benoit Mandelbrot coined the term Fractal,
and described it as follows:

> A [fractal is a] rough or fragmented geometric shape that can be subdivided into parts,
> each of which is (at least approximately) a reduced/size copy of the whole.

He derived the word from the Latin word 'fractus', that implies broken something.
The word 'fractal' is a collective name for a diverse class of geometrical objects,
or sets, holding most of, or all of the following properties (Falconer 1990):

1. The set will have fine structure, it will have details on arbitrary scales.
2. The set is too irregular to be described with classical Euclidean geometry, both
   locally and globally.
3. The set has some form of self-similarity; this self-similarity does not necessarily need to be exact (viz. geometric self-similarity), it can be approximate or
   statistical self-similarity also.
4. The Hausdorff dimension of the set is strictly greater than its topological
   dimension.
5. The set has a simple definition, i.e., it can be defined recursively.

– The property (6) is Mandelbrot's original definition of a fractal; however, this property has been proven not to hold for many sets that are actually fractal. Hence, Mandelbrot's definition of fractals is no longer considered to be the most complete definition of fractals. In fact, each of the five features mentioned above have been proven not to hold for at least one fractal set. Several attempts have been made hitherto to construct a pure mathematical definition of fractals, but all have been proven unsatisfactory. We will, therefore, rather loosely, use the above properties throughout this book when talking about fractals (Falconer 1990).

The essential property of a fractal is that its structure appears to be the same when it is examined on any scale of magnification. Though these words are easy to say or write, they imply a somewhat peculiar property, called self-similarity. There is a reason why I claimed it to be somewhat peculiar. Think of it, when looking out for the finer details of a structure, a kid will always love to examine it under a magnifying glass, a grown-up will do the same under a microscope. We do so because the usefulness of these magnifying devices lie in the fact that they make magnified structures look different from that seen by the naked eye. However, while observing a fractal, these devices will fail! Fractals embody the same level of complexity at all the level of magnification. What is even more peculiar is the fact that Nature has organized itself with very many types of fractal objects. Why does Nature prefer constructing all sorts of fractal objects to run natural organizations? And that too, *so* consistently?—These are deep and difficult questions that involve exploring the interfaces of thermodynamics, geometry and many other fields. No satisfactory, general answer to these questions has been found till now. In fact, truth be spoken, not many scientists dare to ask these questions, let alone attempting to find an answer for them. All that is known is that self-similarly and self-organized structures in nature develop spontaneously from dynamical processes involved in structure formation. The displayed self-similarity can loosely be considered as a critical manifestation of the time-dependent and context-dependent 'tug-of-war' between two antagonistic properties, order, and disorder. To what extent the complexity of nature is attributable to accidental randomness—is not known to us, as of now. Thus, to what extent the aforementioned 'tug-of-war' is equivalent to (confrontation-filled) coexistence of both determinism and accidental development—isn't clear either. In other words, although we have learnt (quite) a bit about the schemes of description of self-similar systems and self-similar processes, we haven't made much headway to understand the general reasons why nature favors them so much. (Interested readers can have a look at (Schroeder 1991) and (Peitgen et al. 1992)— for some helpful (but largely inconclusive) discussions on this line.) Anyway, remaining truthful to the prevailing scientific culture, this book will not venture into these topics either; instead, it will attempt to talk about somewhat simpler questions involving fractals. We will examine the various types of fractal properties of an object of enormous importance, the proteins. But before than that we need to take a somewhat closer look at the fractals. What they really are?

Well, let us begin at the very beginning; let us attempt to measure the length of a curve. This is a "simple" problem that we routinely come across and apparently, "solve", with ease. Typically, we take a straight ruler of length L and then starting from one end of the curve we walk the ruler along the curve for it's the entire length. In doing so, if it takes total number of N steps to traverse from one end of the curve to the other, we conclude that the total length of the curve is N-times-L. Everyday experience suggests that with a smaller (straight) ruler, it takes more number of steps to cover the length of the curve. In other words, N is proportional to $1/L$. We can comprehend from our experience that for really small rulers, that is, as we go to the limit of l tending to 0, the product $N.L$ will gradually become more and more independent of L and this product is what we term as the exact length of the curve. For example, let us measure the circumference of a circle,—a fairly straightforward problem. We take a circle with radius r and inscribe it in a regular polygon of n sides. Let the length of each side of the polygon be denoted by l. Then, the perimeter of the polygon will come out to be: $n.l$. Remembering our school geometry, we note further that if the chord formed by the side of the polygon subtends an angle θ, then the number of sides of the polygon, viz. n, can be expressed as $n = \frac{2\pi}{\theta}$ and each side l can be expressed as: $l = 2.r.\sin\left(\frac{\theta}{2}\right)$. Therefore, the perimeter of the polygon comes out to be

$n.l = \dfrac{2.\pi.2.r.\sin\left(\frac{\theta}{2}\right)}{\theta} = \dfrac{2.\pi.2.r.\sin\left(\frac{\theta}{2}\right)}{2.\left(\frac{\theta}{2}\right)}$. In the case when l becomes really small, that

is, when limit of l tends to zero, θ also tends to zero and $\sin\left(\theta/2\right)$ tends to simple $\theta/2$.. The rest leaves us with known formula of perimeter of a circle.

Point is that, we could perform all of these because a circle is a "smooth curve". When mathematicians talk of a "smooth curve" they merely do not talk about a smooth looks of the curve, what they *really mean* is that if one describes a curve by $y = f(x)$, then the function $f(x)$ is differentiable everywhere. Why is differentiability, you may ask, such an important criterion?—It is important because thanks to such a condition one can express the y at any point x in the neighborhood of x_0 as $y(x) = y(x_0) + (x - x_0)\left[\frac{df}{dx}\right]_{x_0} +$ higher order terms. Then, for a sufficiently small interval of $(x-x_0)$, one ends up with $y(x) = y(x_0) + (x - x_0)\left[\frac{df}{dx}\right]_{x_0}$ which is nothing but a straight line. Thus, a curve can be approximated as a straight line at a sufficiently small interval in x, *only* if it is a "smooth curve" at that point and its neighborhood. This is the reason why, at a sufficiently small value of l, a circle could be represented with a polygon.

But what happens to the cases when such nice limits do not exist? That is, instead of being "smooth", if a curve appear rough and wiggly? Furthermore, try to envisage a situation when the "wiggly"ness of this curve persists no matter how closely you look at it. Surely, the tool-set put on display beforehand, won't find much of a use then,—right? This is where we start to think about functions that are continuous at every point but nowhere differentiable, in other words, we start thinking about fractals and start to appreciate the brilliance and depth of Mandelbrot's works, something that has been talked about earlier in this chapter.

1.3 Misconceptions About Fractals

Although the discussion of above introduces us to the idea of fractals, we should be guarding ourselves from some misconceptions about fractals. This job is far from being simple because misconceptions galore in the field of fractals. Here we talk about some very common misconceptions. For example:

Misconception 1: *Anything that is self-similar is necessarily fractal.* This is, of course, not true. A straight line is a self-similar structure. Any small part of a straight line looks exactly like a large stretch of a straight line, which, in turn, looks like the whole straight line. But that does not make a straight line to be a fractal object. Many think that the reason for straight line's not being a fractal has got to do with the fact that for a straight line, the topological dimension equals the Hausdorff dimension. While this is not wrong, such a reason appears to be a bit mechanical and does not always present the physical picture. Rather, observe the fact that for a straight line, the reduction factor can be arbitrary; while for fractal object, the reduction factor needs to be characteristic. In other words, one needs to follow a rule (for recursion), which ultimately produces the necessary reduction required to construct a fractal object. For example, the Koch curve, described earlier, can be reduced only by factors of $\frac{1}{3^n}$ (where n is integral), to obtain the self-similarity. For straight line, such a definite rule for recursion does not exist. Hence, do not think that all self-similar structures are fractal structures.

Misconception 2: *Fractal dimension must always be fractional.* It comes as surprise, but this is not true either. The Peano curve (mentioned earlier) and some other fractals (the Devil's staircase, for example) have integer dimensions. For the Peano curve (which is a space-filling curve, that starts from a 1 dimensional straight line but applying a definite rule, gradually fills up the entire space, viz. a 2 dimensional area), the fractal dimension is 2.

Misconception 3: *A bounded curve having infinite length must be a fractal.* This, again, is a wrong idea. A Koch curve is bounded (that is, a Koch curve is confined within a finite region of plane) and it has infinite length. But all the curves that are bounded and that have infinite lengths may not be fractal. Take, for example, the case of a spiral. (Diagram of a simple spiral is provided in Fig. 1.10). Such a spiral is made up of circle segments of radii r_k, so that the arc length s_k of the kth circle segment comes out to be $\left(\frac{\pi}{2}\right) r_k$. Hence the length of the spiral is given by: $l = \sum_{k=1}^{\infty} s_k = \left(\frac{\pi}{2}\right) \sum_{k=1}^{\infty} r_k$—but this expression diverges at $r_k = \frac{1}{k}$.

To clear up these conceptions and many others, interested readers can benefit much from studying (Feder 1988); (Falconer 1990) and (Peitgen et al. 1992).

Another important point (rather, another huge misconception) should be clarified here. The present discussion on the background of present-day fractals, should not be confused with the history of fractional calculus, which has an older but much smoother history. On September 30, 1695 L'Hospital wrote a (famous) letter to Leibniz asking him about a particular notation that he had used in his

Fig. 1.10   A simple *spiral*

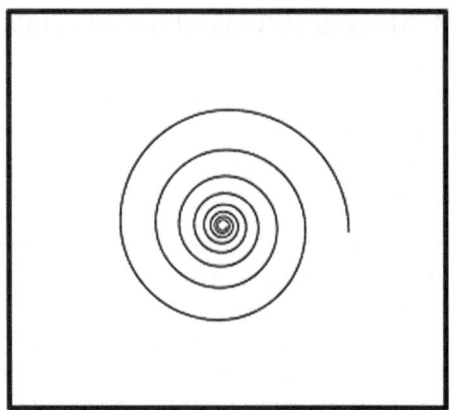

publications for the nth-derivative of the linear function $f(x) = x$, $\frac{D^n x}{D x^n}$. L'Hospital's question to Leibniz was simple—what would be the result if n equals 1/2. Leibniz's response to this question was farsighted: "(this is).. an apparent paradox, from which one day useful consequences will be drawn". After this, unlike the course of historical development of 'continuous-everywhere-but-differentiable-nowhere' cases, many of the best minds of the then-contemporary mathematics (Fourier, Euler, Laplace, to name a few)—have worked on this problem. Hence, the crowning label of being "pathological" was not attached to works that searched for possible implications of cases with noninteger exponents. Just like the idea of coming across a case that is 'continuous at every point but differentiable nowhere'—seems strange for the first time, the contrast obtained from the difference in conceiving the physical ideas for x^2(that is, x multiplied by another copy of x) and $x^{2.37}$ (that is, x is multiplied by 1.37 copies of itself)—seems a bit odd. But then, the very fact that (nonpathological) calculus could routinely work with x^π throughout the whole nineteenth century, suggests that historical evolution of fractional calculus was not as rough as that in the case of fractals. [Interested readers can refer to (Podlubny 1999) and (Nishimoto 1991) to learn more about evolution and various aspects of fractional calculus. To explore the interface between fractional calculus and fractals, one can refer to (Bruce and West 1999)].

Coming back to fractals, although the mathematical properties of fractals have been studied for over 150 years (as we have talked before), fractals were a rather isolated field of mathematics until the 1970s, when, as a result of the efforts of Mandelbrot and others, many people became convinced that fractals had a number of important applications in physical and biological science that range from Brownian motion to chemical reactions. Such subdued nature of initial reactions appears a bit strange, at least from today's perspective, because the trust in the symmetry in description of natural processes and deciphering the cause of natural processes—has been a guiding principle in scientific pursuits for ages. Studies that banked on symmetry under mirror reflection, rotation, translation, and transformation of co-ordinates in general are considered as the cornerstones of scientific understanding of the dynamics of natural processes. But the symmetry of self-similarity had remained hidden from the eyes of the scientists until the computers came to picture. Notably, scientists always associated symmetry with the idea of a well-ordered nature, smooth, and regular. The scientific idea of beauty also revolved around these concepts of symmetries that could make objects move, grow, or shrink following an orderly arrangement in space and time. In contradiction to this school of understanding of symmetries, self-similarity, that left properties of the systems invariant under the change of sizes, revealed very irregular, complex, and chaotic appearance. Predictably, therefore, it took a while before everyone could appreciate the "unusual" nature of the newfound symmetry measure. Only recently it has been established that many (but not all) of such irregular and apparently chaotic systems can be described with the help of symmetry of self-similarity.

The "subdued reaction" talked about in the last paragraph, portrayed some kind of rigidity in scientific culture. When viewed from present scenario, such rigid stances appear to be strange (if not bizarre). That is because most systems around us in nature—trees, rivers, rocks, mountains, clouds, fire, sand-beds, coastlines, etc.—are irregular and unsmooth and they were always irregular and unsmooth. Mandelbrot's understanding of nature prompted him to think that a fractal would mathematically model a cloud better than a sphere would. The question of "How long is the Coast of Britain" may seem trivial, but when one considers the scale of our measurement—the question appears far from being simple. From a satellite orbiting around the earth in low altitude, one can make a rough approximation; but for an ant walking the shore, following every sub-inlet, every creek and every stream as it runs deep into the British Isle, the measured length tends toward infinity. Characterizing the coastline as having a fractal dimension of 1.2, therefore, proved to be a more accurate answer to the aforementioned question. By the early 1990s, it became clear that fractal geometry is a powerful tool to describe most of the structures we encounter in our everyday life. From this point onward, various works during the last three decades have portrayed a totally different picture. Since fractal systems have a self-similar (or at times, self affine) geometry at different measurement scales, results obtained from measurements made at one scale can be used to predict geometries at other scales. This property of scale-independence has been recognized as a very useful one and is therefore used extensively in studying numerous types of natural phenomena.

While the developments in fractal geometry were taking place, similar developments were taking place in a new branch of Physics, the chaos theory. The two ideas fed each other, sharing similar roots mathematically. Thanks to the simultaneous growth and symbiosis of chaos theory and fractal geometry, the idea of a clockwork universe (a universe where a knowledge of all of the initial states of all particles and exact knowledge of the laws of physics would allow you to calculate the evolution of the universe from that point onwards exactly)—was brought down entirely.

Why are we studying proteins?—Well, apart from their decisive roles in performing biological functions, proteins present themselves as qualified objects to study the dynamics of complex systems. Because of the aperiodic arrangement of protein constituents, viz. the amino acids, conformational states of proteins consist of many sub-states with nearly the same energy. Thus, potential energy surface of a protein, almost invariably, turns out to be a rough hyper-surface in a high dimensional configuration space with an immense number of local minima. Proteins share this feature with other complex systems like spin glasses, glass-forming liquids, macromolecular melts, etc. But an advantageous aspect with proteins is that, compared to other (aforementioned) systems, a protein is a relatively small (and therefore manageable) system to work with. Furthermore, proteins are identically reproduced by nature (that is, one less worry to overcome). Therefore in recent years, proteins, especially small proteins like myoglobin, have become model systems to study dynamics of complex systems.

Throughout the course of this book we will talk about the (incontrovertible) relevance and suitability of studying protein interior with fractal dimension-based measures.

References

Besicovitch AS (1929) On linear sets of points of fractional dimensions. Math Ann 101(1):161–193. doi:10.1007/BF01454831

Besicovitch AS, Ursell HD (1937) Sets of fractional dimensions. J London Math Soc 12(1):18–25

Bourke hypertext link (2012) http://paulbourke.net/fractals/fracdim/. Accessed 27 Sept 2012

Boyer CB, Merzbach UC (1989) A history of mathematics, 2nd edn. Wiley, New York

Butzer PL, Stark EL (1986) "Riemann's Example" of a continuous nondifferentiable function in the light of two letters (1865) of Christoel to Prym. Bull Soc Math Belgique 38:45–73

Cantor G (1883) "Über unendliche, lineare Punktmannigfaltigkeiten V" [On infinite, linear point-manifolds (sets)]. Math Ann 21:545–591

Cellérier C (1890) "Note sur les principes fondamentaux de l'analyse" (Note on the fundamental principles of analysis). Bull des Sci Math 14:142–160

Edgar GA (1990) Measure, topology, and fractal geometry, 1st edn. Springer, Berlin. ISBN 0-38797-272-2

Falconer KJ (1990) Fractal geometry: mathematical foundations and applications. Wiley, New York. ISBN 0-471-92287-0

Feder J (1988) Fractals. Plenum Press, New York

Folta J (1966) Bernard Bolzano and the foundations of geometry. Acta Spec Issue 2:75–104

Gleick J (1987) Chaos: making a new science. Penguin, New York

Hausdorff F (1919) Dimension und "ausseres mass. Math Ann 79:157–179

Hilbert D (1891) Ueber die stetige Abbildung einer Line auf ein Flächenstück. Math Ann 38:459–460. doi:10.1007/BF01199431

Jasek M (1922) "Funkce Bolzanova" (Bolzano's function). Casopis pro Pestování Matematiky a Fyziky (J Cultivation Math Phys), 51(2):69–76 (in Czech and German)

Kaye BH (1994) A random walk through fractal dimensions, 2nd edn. VCH, Wiley. ISBN 3-527-29078-8

MacTutor (2000) Hypertext file: http://turnbull.mcs.st-and.ac.uk/history/. Accessed 27 Sept 2012

Mandelbrot BB (1967) How long is the coast of Britain? Statistical self-similarity and fractional dimension. Science 156:636–638

Mandelbrot BB (1975) Les objets fractals, forme, hasard et dimension. Flammarion, Paris

Mandelbrot BB (1982) The fractal geometry of nature, 1st edn. W. H. Freeman and Company, San Francisco. ISBN 0-7167-1186-9

Marstrand JM (1954) Some fundamental geometrical properties of plane sets of fractional dimensions. Proc London Math Soc 4(3):257–302

Mumford D, Series C, Wright D (2002) Indra's pearls: the vision of Felix Klein. Cambridge University Press, Cambridge

Neuenschwander E (1978) Riemann's example of a continuous 'nondifferentiable' function. Math Intelligencer 1:40–44

Nishimoto K (1991) An essence of Nishimoto's fractional calculus, Descartes Press Co., Koriyama

Nový L (1981) Poznámky o "stylu" Bolzanova matematického myslení [Remarks on the "Style" of Bolzano's Mathematical Thinking], DVT 81:217–227 [Czech, English and Russian summary]

Peano G (1890) Sur une courbe, qui remplit toute une aire plane. Math Ann 36(1):157–160. doi: 10.1007/BF01199438

Peitgen H-O, JÄurgens H, Saupe D (1992) Fractals and chaos: new frontiers of science. Springer, New York

Podlubny I (1999) Fractional differential equations, "mathematics in science and engineering V198", Academic Press, Waltham

Richardson LF (1961) The problem of contiguity: an appendix to Statistic of Deadly quarrels. General systems: yearbook of the society for the advancement of general systems theory. Ann Arbor, Mich: Soc (1956) Soc Gen Syst Res 6(139):139–187

Rocco Andrea, West BJ (1999) Fractional calculus and the evolution of fractal phenomena. Physica A 265:535–546

Chapter 2
Studying Protein Interior with Fractal Dimension

Abstract Here we look at the interior of protein structures, as viewed through the prism of fractal dimensions. History of application of fractal dimension-based constructs to probe protein interior dates back to the development of theory of fractal dimension itself. Numerous approaches have been tried and tested over a course of 30 years. All of these bring to light some or the other facet of symmetry of self-similarity prevalent in protein interior. Later half of the last decade, especially, has been phenomenal; in terms of the works that innovatively stretched the limits of fractal dimension-based studies to present an array of unexpected results about biophysical properties of protein interior. Here, starting from the fundamental of fractals, we will learn about the commonality (and the lack of it) between various approaches, before exploring the patterns in the results that they have produced. Clustering these approaches in major schools of protein self-similarity studies, we will attempt to describe the evolution of fractal dimension-based methodologies. Then, after attempting to work out the genealogy of approaches (and results), we will learn about certain inherent limitations of fractal constructs. Finally, we will attempt to identify the areas and specific questions where the implementation of fractal dimension-based constructs can be of paramount help to unearth latent information about protein structural properties.

2.1 Why, at All, Does One Need Fractal Dimension to Study Protein Interior?

Anybody who has worked on any small little problem on anything to do with protein structure will tell you that proteins are rather complex and enigmatic entities. You may ask, "precisely how much enigmatic?"—Well, to present an idea, think of "simple" single domain globular proteins. They undergo remarkably cooperative transitions from an ensemble of unfolded states to well-ordered folded states (often called, somewhat loosely, the 'native states') as the temperature is lowered. Protein scientists, thousands of them, from all across the world, have accumulated enough materials to describe the various stages of the aforementioned cooperative transition studying various proteins. But, despite intense research of more than half-a-century, scientists could not decipher the general principles how proteins fold with utmost reliability and consistency to their native conformations, despite frustration in the form of nonnative interactions between residues. What makes this open question seem tantalizingly close to our reach is the fact that in many cases, the transition to the native state takes place in an (apparent) straightforward two-state manner; i.e., the only detectable species are found to belong to either to the unfolded state or to the native state (Poland and Scheraga 1970; Creighton 1993; Privalov 1979). Studies have shown that folded states of globular proteins are only marginally stable below the folding transition temperature and the free energies of stability of the native state, relative to the unfolded states, vary within a small range of (5–20) k_BT at neutral pH (Creighton, 1993; Taverna and Goldstein 2002). Furthermore, it has been identified that the transition to the native state at the folding temperature has the characteristics of a first-order phase transition (Privalov 1979; Bryngelson et al. 1995; Kolinski et al. 1996; Dill et al. 1995; Li et al. 2004). That is because the transition to native state at the folding temperature could be categorized as one due to effective inter-residue attractions that compensates for the loss of entropy. However, we have not succeeded in fully understanding the microscopic origin of this inter-residue cooperativity (Bryngelson et al. 1995; Kolinski et al. 1996; Dill et al. 1995; Li et al. 2004), even in the case of "simple" small single domain globular proteins. This is why, although it seems tantalizingly close, we are yet to understand the structure and folding aspects of proteins, fully. Thus, there is reasonable ground to assert that proteins are enigmatic characters, full of nonidealistic traits that are farthest from being addressed with simplistic constructs. Not surprisingly therefore, one work has described proteins as 'complex mesoscopic systems' (Karplus 2000).

Apart from being characterized by an enormous number of degrees of freedom, proteins have multidimensional potential energy surfaces. It is a known fact (De Leeuw et al. 2009; Reuveni 2008) that while a protein needs to adhere to its specific native fold to ensure its (thermal) stability, this native fold should allow certain parts of the protein to undergo large-amplitude movements. These movements (studied under the broad umbrella of fluctuation studies) assume a critical character in ensuring the desired functionality of a protein. It is equally intriguing to notice that although the structure of a protein can be accurately studied as that of a composite macromolecule, it has many a properties in common with that of macroscopic microparticles (Goetze and Brickmann 1992). From somewhat a different perspective, crystallography studies (Tissen et al. 1994) have reported on the mesoscopic nature of protein structures. The standard 'compact object description' of proteins (characterized by small-amplitude vibrations and by a low-frequency Debye density of states) cannot account for their 'non-idealistic' behavior (De Leeuw et al. 2009; Reuveni 2008). Indeed, the nonconstancy of distance between any two atoms $\left(\left| \vec{r}_i - \vec{r}_j \right| \neq \text{Constant} \right)$ in any biologically functional protein can easily be verified with the simplest of computer programs. In addition, recent (Banerji and Ghosh 2009a, 2011) characterizations of inhomogeneous distributions of mass and hydrophobicity in the protein interior merely serve to complicate an effort to construct a straightforward and linear scheme for the description of the protein interior.

Thus, it is unrealistic to expect that constructs that are otherwise adequate to describe the complexity of simple structures (spheres, cubes, other regular structures of idealistic shape, and characteristics) will help in attempts to describe proteins. Clearly, to achieve this goal one needs to think beyond the tool set of simple geometry and the physics of idealistic systems. To be categorical, one needs to adopt an approach that has the capabilities to describe the inherently inhomogeneous and nonlinear behaviors of protein structural parameters as an innate capacity of proteins, rather than treating them as deviations from some arbitrarily chosen standard for (illusory) ideality. In this very context, one notes that quantification of the self-similarity prevalent in protein properties can serve as a potent tool to achieve the aforementioned challenging task. Self-similarity has a unique advantage over many other possible constructs, because an objective evaluation of self-similarity will enable deciphering of the hidden symmetry that connects global patterns of macroscopic properties in proteins (such as hydrophobicity distribution, polarizability distribution, etc.) with the local (atomic) interactions that produce them.

During attempts to describe many natural phenomena, researchers found that fractal dimension (denoted as FD from here onward) could reliably quantify self-similarity. Hence, resorting to FD to quantify the self-similarity prevailing in the distributions of protein biophysical and biochemical properties does not appear to be an unreasonable proposition. As a result, investigations of protein structures using FD-based measures have generated enormous interest.

2.2 Schools of Protein Interior Fractal Studies

Two forms of protein information, viz. one of the primary structures, and that of the three-dimensional structures of proteins (Stapleton et al. 1980; Havlin and Ben-Avraham 1982a, b, c, d; Alexander and Orbach 1982; Wagner et al. 1985), were chosen to conduct preliminary investigations. The school of primary structure studies owes its origin to polymer research primarily and focuses for the most part upon examining fractal properties within backbone connectivity. The other school of studies largely ignored main-chain information and concentrated on the distribution of residues and atoms within protein space. These schools, together, constructed the general framework with which one can quantify interior fractals. Various sub-approaches can be observed within both these broad-based approaches that differ from each other (mainly) in terms of their implementation strategies. There are also two other basic approaches—slightly more abstract—that have been used to probe the protein interior. One of these revolves around the application of correlation dimension-based constructs to explore interior structural invariants, the other attempts to relate renormalization group theory (RGT) in the realm of protein structure analysis. It is necessary to (briefly) survey these lines of thoughts when comparing and contrasting protein structure. Figure 2.1 presents a schematic view of the families of FD-based approaches.

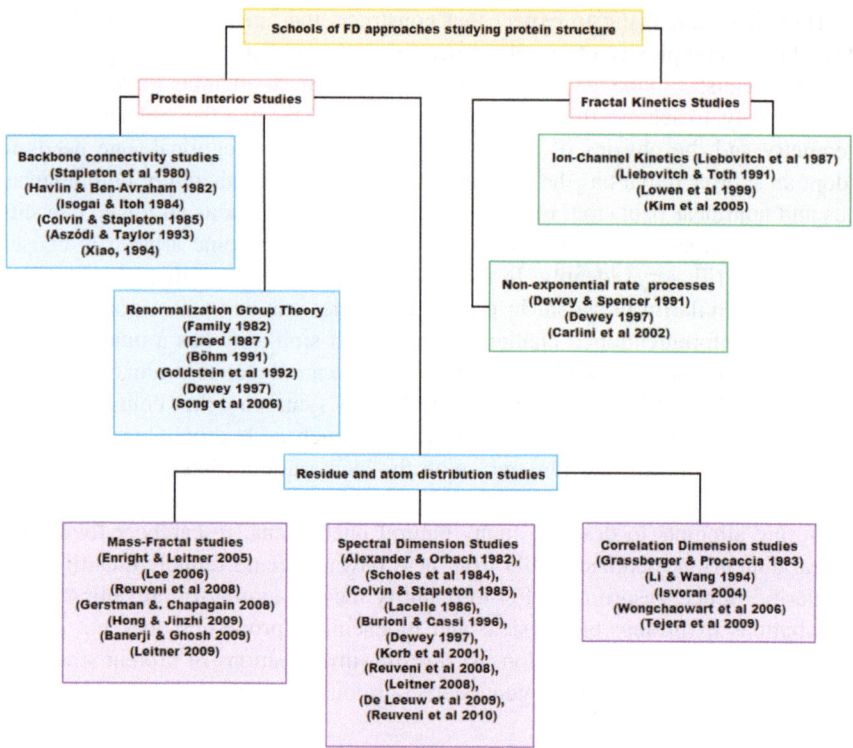

Fig. 2.1 Schools of protein interior fractal studies

2.2.1 Approaches with Backbone Connectivity

Primary investigations in these lines were started by Havlin and Ben-Avraham (1982a, b, c, d). Their studies, however, did not directly involve much of proteins. But these pioneering investigations paved the way for many a works during the next three decades that attempted to study protein primary structures with fractal dimensions. Amongst them, the work 'Fractal dimensionality of polymer chains' (Havlin and Ben-Avraham 1982a) had a phenomenal impact on the early works on protein fractal studies. Thus, it assumes significance to understand it. This work was important because it was (probably) the very first study that asserted that FD could be a reliable marker while investigating configurational properties of a single polymer. Make no mistake, the various statistical configurational properties of long polymer chains were known already, both from theoretical as well as experimental perspective (de Gennes 1979; Flory 1971). Behaviors of several properties of statistical ensemble of polymers were known too (McKenzie 1976). But, due to some unknown reason, properties of a single configurational shape of a single regular polymer in statistical equilibrium were not known till then. Properties those were otherwise adequate, like mean radius of gyration (ROG), mean square end-to-end distance, etc.—were found lacking for the aforementioned job. The authors (Havlin and Ben-Avraham) asked: what happens to the case when a regular polymer chain does not change its general shape because of bending in the middle? For such a bending, the mean square end-to-end distance will change and so will the mean ROG. It is in this context that the use of FD was proposed to characterize the configurational properties of a single polymer. Upon modeling the polymer by long self-avoiding walk (SAW), the FD was calculated. In terms of magnitude, the obtained FD did not differ much from the average value of the entire ensemble of polymers; however, conceptually, it was a major leap.

The three papers of Havlin and Ben-Avraham (1982a, b, c), all published in 1982, studied the fractal properties of polymer chains and thereby laid down the ground for subsequent treatments of similar type on proteins. Denoting the average end-to-end length of an unbranched polymer by L, the number of monomer segments (of length ε) by N, and fractal dimension as FD, they found the scaling relationship to follow: $N = \left(L/\varepsilon\right)^{1/v_F} = \left(L/\varepsilon\right)^{FD}$, where the exponent turns out to be the inverse Flory constant v_F in polymer theory. [Theoretical considerations provide limits $1 < FD < 2$, which corresponds to a linear structure given by: $L = \varepsilon N$ on one end, and a structure represented by unrestricted random walk, when FD approaches the topological dimension magnitude 2, where $L = \varepsilon N^{1/2}$ (Tanford 1961), on the other].

Two similar approaches emerged from this basic platform. Both considered protein length to be the most important variable, but they differed on the exact definition of protein length. The first method attempted to find the scaling exponent of the contour length, which is proportional to the number of monomers with respect to the end-to-end length. The second method attempted to quantify the length of a line segment along the chain, measured with a scale of the number of monomers in the segment. Because of its algorithmic strategy, the second method is alternatively denoted as 'the internal line segment method'. But to appreciate the intricacies of these algorithms one needs to acquire a little idea about a construct named SAW at first.

A SAW is a sequence of "walks" on a lattice, where the walk does not visit any point for more than once. In other words, while "walking" from one point to another, a SAW never intersects itself. For wonderful pictorial description of them please visit—Wolfram Web Resource on SAW (Weisstein, SAW, hypertext link: http://mathworld.wolfram.com/Self-AvoidingWalk.html). There is nothing sacred about using the lattices, but it so happens that calculations on lattices offer certain advantages (viz., easier length measurement, control on the step direction, discreet number of steps etc.) without sacrificing the generality;—hence lattices are considered as model systems to study the properties of SAW. Those with slight background in this field will be immediately able to identify that SAW is a lattice random walk (or 'Polya walks', as it was introduced by George Polya), with one additional constraint, viz. no point can be revisited. For a lucid nonmathematical tour of SAW one may prefer the article 'How to Avoid Yourself' by Hayes (1998); more mathematically inclined readers, in addition, can refer to the old classics

(Kesten 1963) and (Domb 1969). SAWs have intrinsic mathematical interest, and their study involves an astonishingly large range of coverage in physics, chemistry, and biology. Polymer chemists often take recourse to SAWs to model the properties of polymers, whose physical volume imposes the constraint that multiple occupation of the same spatial point is not possible. It is helpful for polymer modeling for other reasons too. Owing to the forbidden self-intersections, SAW path corresponds to the excluded-volume interactions of monomers that comprise the polymer chain. SAW-based modeling has been immensely fruitful in the paradigm of "knotted-structures" like proteins too. Importantly, in the context of present discussion, SAWs have been shown to have fractal properties (Havlin and Ben-Avraham 1982b, c, d); when considered in Euclidean dimension = 2, the FD is found to be 4/3, while for Euclidean dimension = 3, the FD is ~5/3. For an interesting study on the extension of these ideas, one can refer to (d'Auriac and Rammal 1984).

Now, armed with this (elementary) knowledge of SAW we can revert back to the problem of calculating the fractal dimension of protein backbone. As mentioned previously, based on the basic platform laid out by Havlin and Ben-Avraham's, two similar approaches emerged. Both considered protein length to be the most important variable, but they differed on the exact definition of protein length. In the first method (Havlin and Ben-Avraham 1982a, b; Wagner et al. 1985; Colvin and Stapleton 1985), protein length was defined as:

$$\langle Z\,(n)^2 = \frac{1}{N-n+1} \sum_{i=1}^{N-n+1} Z_{i,i+n}^2 \rangle$$

where $Z_{i,i+n}^2$ is the square of distance between the terminal residues 'i' and '$i+n$', of the chain with N amino acids and n intervals. The FD, in this case, is calculated by performing linear regression in the scale-range appropriate to the logarithmic form of: $\left[\langle Z(n)^2 \rangle^{1/2}\right]^{FD} = \text{Const}.n$.

The second method, although banked on regression for the calculation of FD magnitude, differed from the first one in its definition of protein chain length. This approach (Isogai and Itoh 1984; Wang et al. 1990) defined protein length as: $L(n) = L(m) + \frac{N-n\mu-1}{n\mu}L(m)$

where N is the number of amino acids, n is the total length interval, $L(m)$ corresponds to the length of the chain for the first m-integer segments and $\mu = \left[\text{int}\,(N/n) - 1\right]$. The second term quantifies the remaining $(n-m)$ length of the segment. Qualitatively it implies that with this scheme the emphasis rests primarily on quantifying the scaling behavior of the segmental length of the backbone, which is defined as the sum of stepwise connections of straight lines, measured for different interval of monomers. Fractal dimension can then be obtained by analyzing $L(n) = \text{Const}.n^{1-FD}$, only in the scale range where $\ln\,(L(n))$ and $\ln(n)$ are linearly related. Such linearity in the log–log plot corresponds to the scale invariance associated with the observation of the phenomenon under consideration. Some 10 years after this methodology was published, Xiao (1994) suggested a modified version of it by proposing two possible correction schemes. These were achieved by implementing the end-to-end distance of the remaining residues and by performing $L(n)$ measurement starting from different C^α atoms, instead of terminal C^α only. It is observed that these corrections increase the quality of the regression for high values of n, probably because it is only then the statistical nature of the process becomes more apparent. A detailed comparison of this genre of FD-based measures with the others that attempt to describe symmetry in protein backbone chain can be found in following references (Bytautas et al. 2000; Aszodi and Taylor 1993).

FD magnitudes obtained from first and second methods, in the case of SAW model, were 5/3 and 1.40, respectively (Havlin and Ben-Avraham 1982a, b; Xiao 1994). It was known that, in three dimension, FD would approach the topological dimension 2 for the Gaussian walk (or θ-solvent); whereas FD will assume a magnitude 5/3 for SAW, which includes the excluded volume effect (de Gennes 1979). However, since the SAW does not take into account any attractive force, a deviation from idealized magnitude 5/3 was expected too, especially when there are interactions between monomers that might be far apart along the main chain, but close within folded protein (Elber 1989). Thus, more realistic polymers are likely to have chain FD different from that of the theoretically predicted SAW magnitude, which is precisely what was observed from the obtained FD values. For more comprehensive discussion on FD-based description schemes for polymer collapse and polymer scaling in general, one can refer to (de Gennes 1979; Dewey 1993, 1995). But, since many of the early fractal works on proteins were firmly rooted in polymer studies, it assumes importance here to talk a little on this issue, before we get back to the realm of proteins again.

2.2.1.1 A (Qualitative) Comparison Between Polymer Collapse and Protein Folding

Early methods of calculating the fractal dimension of any protein depended principally on primary structure information, but the methodological apparatus for these investigations were (largely) derived experiences from polymer research. Algorithms presented beforehand and the discussions therein, demonstrate that. Although proteins are special types of polymers and although polymer–centric approaches are undeniably associated with protein fractal studies, there are differences between proteins and polymers in general. Therefore, before we proceed any further, it is important that we briefly compare and contrast two similar phenomena, polymer collapse and protein folding. But just like the fact that proteins are characterized by a large number of degrees of freedom, field of protein folding studies is enormously multifaceted too. Since it is difficult to encompass the entire array of aspects related to polymer collapse and protein folding, we will restrict ourselves to discussions related to some (reasonably) direct questions.

One may ask, why, at the first place, protein folding studies are bothered about treatments of polymer collapse? Probably, it is because of the hope that polymer collapse-type treatment of protein folding problem may help in identifying the general principles that govern the later. To elaborate a little, one needs to identify the exact universal (critical) scaling properties of proteins in order to generalize the principles of protein folding to all the proteins without bothering about whether it is protein 'A' or protein 'B', in context A* or context B*. For such a task, the statistical mechanics based models of conformations of linear polymers in dilute solution in two or three dimensions—come handy; whereby an extension of polymer collapse models appear to be rational. Or else, one may choose to look at the question (viz., 'why to study polymer collapse in order to study protein folding?') from a historical perspective also. Perhaps the most fundamental of all the properties of an isolated polymer chain is its size and shape under various solvent conditions (de Gennes 1979; Doi and Edwards 1988; Grosberg and Khokhlov 1994). Since, in the early days of protein structure studies, the primary structures were more easily available than the tertiary or quaternary structures; and since, the widely held belief was that the spontaneous folding of amino acids (protein monomers) into unique globular shape is *solely* determined by the sequence of amino acids itself, it was not unusual to resort to polymer collapse models while attempting to describe protein folding.

What is the basic premise of polymer models? For completely flexible chains in a "good" solvent, the monomers of polymers will favor contacts with the solvent. This phenomenon can alternatively be described by saying that in good solvents, monomers effectively repel one another. As a result, in good solvent, the polymer shape resembles that of a swollen coil. In "poor" solvent, in contrast to the case of the good solvents, monomers are repelled by the solvent. Speaking alternatively, in poor solvents, the monomers attract one another, and an isolated chain, therefore, forms a compact globule. Such a globular form can be (conveniently) assumed as spherical in shape, which minimizes the surface area between monomers and solvent. Barring these two extreme cases, one finds other polymers that exhibit substantial bending stiffness, and therefore find it difficult to attain a compact globule shape, because that requires substantial bending. The later cases are exemplified by many biopolymers, such as F-actin and DNA. Such chains are described by the worm-like chain models (Grosberg and Khokhlov 1994).

One can establish the common ground between polymer collapse and protein folding studies, merely based on these elementary facts. When an unfolded polymer chain is transferred into a poor solvent, it collapses to form a compact globule. Similarly, an unfolded polypeptide that is transferred from a denaturing solvent (e.g., guanidine hydrochloride solution) into water, also collapses to a compact configuration. However, the collapsed state does not necessarily imply the existence of the folded state of a biologically functional polypeptide. Hence, one should be careful not to use the word "native" state (denoting the folded state of a biologically functional macromolecule) in place of the word "collapsed" state. The compact denatured configuration, obtained from collapse, has been variously named as "molten globule", the "compact globule", or the "folding intermediate". Since, studies related to collapse transition and the nature of these compact denatured states both lay down the foundation of biophysical studies of protein folding, understanding polymer collapse assumes importance while attempting to understand protein folding.

The (aforementioned) statistical mechanical models display three different behaviors. At high temperatures these models imitate polymers in "good" solvent, which means a solvent where the excluded volume effect is dominant. In contrast to this, in low temperatures, the models describe a collapsed state, caused by solvent-mediated attractive interactions amongst different stretches of a polymer. In one particular temperature, known as the θ-point, the excluded volume effect and attractive forces "balance" each other, to give rise to the third type of behavior. This third behavior is routinely modeled with various variants of random walk models. But it was realized during the last two decades of last century that, the aforementioned paradigm of simplistic and idealistic expectations might not be apt to model protein folding. The ground for this realization was found from the finding that in two dimensions, the θ-state does not share the scaling features expected from a random walk (Duplantier and Saleur 1987). But interesting as it is, since the discourse on this topic deviates significantly from protein reality beyond this point, we won't talk about it anymore. Suffice to note that polymer collapse theories attempt to model how precisely the collapse transition of a single polymer chain happens, in all its aspects. Since it is largely accepted (Lifshitz et al. 1978; Chan and Dill 1991; Pande et al. 1997) that deciphering the nature of various parameters during polymer collapse is foremost a precursor to deciphering protein folding, we will zoom in on only this aspect.

How does polymer collapse take place? The earliest known consistent theoretical framework of polymer collapse was (probably) put forward by de Gennes (de Gennes 1985). According to this framework, as the temperature changes by ΔT from the θ-condition, a Gaussian coil starts to aggregate, forming a uniformly dense sausage-like shape. With passage of time, due to the requirement to minimize the interfacial area, this sausage-like shape thickens and shortens; until, at the final stage, a globule is formed. Alas, if only life were so simple. For various questions of equilibrium statistical mechanics one finds neat mathematical schemes that fit well and fit easily with intuitive expectation. However, matching the intuition does not necessarily imply the infallibility of the scheme itself. This could be observed with de Gennes model.

From the mid-1990s, a barrage of works (Ostrovsky and Bar-Yam 1994; Byrne et al. 1995; Chu et al. 1995; Buguin et al. 1996; Dawson et al. 1997; Halperin and Goldbart 2000) started to report that such uniform sausage-like shapes are highly unstable in solvents owing to the capillary instability; whereby, collapse proceeds by the formation of "pearl necklace" and the gradual diffusion of large pearls from the chain ends. Subsequently, many efforts have been made to incorporate the capillary instability in the description of the collapse of a polymer chain (Klushin 1998; Halperin and Goldbart 2000; Abrams et al. 2002); however, obtaining a universally accepted theoretical description for this system remained a difficult problem. This merely goes to underline the point that the mere construction of a simplified model to describe protein folding may not necessarily be the correct solution.

In this context, it is important to point out certain crucial differences in 'polymer collapse-vs-protein folding' paradigm; alongside these we jot down certain major difficulties. There are many differences and difficulties. We'll start by reminding ourselves certain points about the most important aspect of this discussion, the biological aspect. We know that various types of proteins are meant to function in various types of cellular environments. The 'fibrous proteins' form large aggregates that are (almost) water free. Fibrous protein structures are known to possess high content of hydrogen bonds and they are known to be highly regular and noncompact (owing to separate chains). 'Membrane proteins' reside within the water-lacking membranes. Their intramembrane portions are also highly regular and highly hydrogen bonded, but restricted in size by the membrane thickness (~40 Å). In contrast, the water-soluble 'globular proteins' are less regular (especially at the protein/water interface). In them, the compact interior structure is maintained by interactions of the chain with itself and sometimes with various other molecules (the co-factors). While it may be a physicist's delight to propose an accurate and consistent theory to explain the folding process of all three types of proteins with a single model, it seems less realistic. A collapse or folding theory that addresses the typical characteristics of the three aforementioned types of native states is difficult to find.

Knowledge about the process of protein synthesis suggests that the ribosome synthesizes protein chain residue by residue, from its N-terminal of the chain to C-terminal of the chain. Interestingly, the process of protein synthesis does not take place with uniform rate. There are temporary rests of the synthesis at the rare codons (they correspond to tRNAs which are rare in the cell, and these codons are rare in the cell's mRNAs, too). It is assumed that the pauses may correspond to the boundaries of structural domains that can help maturation of the domain structures. Some enzymes, like prolyl-peptide- or disulfide-isomerases accelerate the in vivo folding. They catalyze the trans↔cis conversions of the backbone conformation of prolines and formation (and decay) of disulfide bonds formed by side chains of cysteines. [Interested readers can refer to (Kolb et al. 1994), (Ellis and Hartl 1999), or (Finkelstein and Ptitsyn 2002) for more details on this (and similar) topics. Anyone is free to guess how many polymer collapse-based protein folding theories attempt to incorporate these biological details in their models.

Next, we focus on the paradigm of theoretical (and computational) comparison between polymer collapse and protein folding. To start with, one should note that proteins are heteropolymers with a variable composition of 20 different amino acids, whereas most of theoretical models tend to resort to homopolymer-based analysis. How much of a difference will that account for? Well, let's consider a problem with appreciable relevance to our present topic, the problem of absorption of an ideal polymer into a point-like potential well. For a homopolymer in equilibrium, one could efficiently characterize the process as second-order phase transition (Binder 1983; Eisenriegler 1993). But, for a heteropolymer, even in equilibrium, the problem has proved to be difficult to be fully understood (Naidenov and Nechaev 2001; Madras and Whittington 2002; Stepanow and Chudnovskiy 2002; Grosberg et al. 2006). [Interested readers may note that an early account of this topic is supposedly provided in (Grosberg and Shakhnovich 1986),—unfortunately this paper is in Russian. That explains why I cannot comment on its content].

Moving on to the next significant difference, by standards of polymer physics, proteins are very small (consisting of merely 50–250 amino acids in general). Such small size is in stark contrast with (statistical) requirements of procedures that are routinely deployed in the sphere of polymer studies. One may choose to remember that the scaling relationships in polymer physics are (principally) derived for large homopolymers.

Lastly, overwhelming body of accumulated evidences tend to suggest that the kinetic ability of natural proteins to fold is acquired, to a significant extent, from the evolutionary selection of the sequence of the amino acids. To what accuracy and consistency, will it be possible for the homopolymer-based analyses to accommodate this concern? But then, one should not be under the misconception that heteropolymer-based modeling (instead of homopolymer-based modeling) is the answer to all the aforementioned problems. There again one needs to ascertain to what extent the model heteropolymers are protein-like? Surely, it is not enough for a polymer to have a unique folded conformation. A heteropolymer with a random sequence will have some lowest energy conformation. Therefore, in conducive temperature and solvent conditions, the polymer will eventually collapse to its folded state,—a fact reported in series of papers in mid and 1990s (Bryngelson and Wolynes 1987; Sfatos et al. 1993; Pande et al. 1995a, b, 1998; Gutin et al. 1996; Wang et al. 1996, 1997). But, probably, there is nothing to be overjoyed at this observation; because the freezing transition of coil-to-folded state of the heteropolymer differs nontrivially from the folding transitions observed in primary structure to native state of the proteins. Typically, in proteins, the folding of random sequences can be characterized as (merely) weakly cooperative (Sfatos et al. 1993; Pande et al. 1995a, b, 1998); furthermore, such a transition in proteins proceeds very slowly due to trapping in metastable conformations that are unrelated to the lowest energy conformation (Bryngelson and Wolynes 1989; Sfatos et al. 1993; Wolynes et al. 1995; Wang et al. 1997). In a related observation, one finds that unlike the case of proteins, the lowest energy conformation of a random sequence of amino acids is expected to be very sensitive to mutations (Shakhnovich and Gutin 1991; Bryngelson 1994; Pande et al. 1995a, b; Bussemaker et al. 1997).

The last finding is significant because it tends to imply that the level of complexity encountered in protein paradigm is far more than that found in polymers. To exemplify this assertion, let us consider the works that attempted to quantify the contributions of secondary and tertiary interactions in influencing folding and stabilizing the native structure of proteins. This is only a small piece in the jigsaw puzzle, but even a small account of this tiny bit illustrates the difference in levels of complexities between two systems. In a series of works in late 1970s, Go and co-workers brought to light the essential nature of specific tertiary interactions in stabilizing the native state (Go and Taketomi 1978; Go et al. 1980). A decade later, Zwanzig and co-workers (Zwanzig et al. 1992) reported that proteins could fold rapidly if there is a rather large local bias toward the correct conformation. But at about the same time Karplus and Shakhnovich (Karplus and Shakhnovich 1992) pointed out, that in such a case, protein folding would not be cooperative and a significant fraction of the population of proteins will not be in the ground state at reasonable temperatures. Within a year, Dill and co-workers (Dill et al. 1993; Thomas and Dill 1993) used simple lattice models with nonlocal interactions and helical propensities to demonstrate that the strength of the local propensities has to be low with respect to that of the tertiary interactions, in order to obtain the average length and number of helices observed in biological protein. While all these results are correct, the apparent contradiction between them originates from the fact that (probably) all of them are incomplete in some sense or the other. The reason for such incompleteness can probably be attributed to the shear extent of complexity found in proteins and not in polymers. The lesson from all the aforementioned reports, therefore, is that one needs to be careful while attempting to apply algorithms and/or arguments that are pertinent in the paradigm of polymer physics to the paradigm of protein folding and protein structure.

But does all of this only suggest that similarities between polymer collapse and protein folding are purely metaphorical? Not at all. The fact that polymer collapse and protein folding are both (at least) partly guided similar processes—stands as the biggest platform of commonality between them. Furthermore, it would have been difficult for protein folding models to explain the Blind Watchmaker's Paradox (viz., how a functional protein can originate from mere random sequences) and Levinthal's Paradox (viz., how the folded state of a protein can be obtained from an astronomical number of possible nonnative states by random search) without applying the polymer collapse models. The sense of generality embodied by the polymer collapse models, therefore, has helped subsequent protein folding research tremendously.

Undoubtedly, the initial contraction of a polypeptide chain from an expanded to a more compact state constitutes the very first stage of protein folding. But then, how exactly to define the premise of the word "initial"? Just like the rate of folding, the rate of the initial hydrophobic collapse can vary over orders of magnitude, depending on the amino acid sequence of the chain. Therefore, the boundary of the word "initial" should be defined in all relevant details. Probably, this is one of the fundamental questions that needs to be answered if polymer collapse-based protein folding models ever wish to make serious impact in the realm of practicing researchers in millions of computational biology laboratories, scattered over all across the world. One may ask many other questions also. Simulations of protein folding in model chains (invariably) result into two extreme cases for the aforementioned "initial" phase (Thirumalai 1995; Gutin et al. 1995; Chahine et al. 2002). Either, as the result of contraction, the computational analog of biological primary structure acquires the folded structure with "specific" collapse, or it contracts rapidly and "nonspecifically" without forming native contacts—and then reorganize into its folded form. How to distinguish between specific and nonspecific collapse? Intuitive approach suggests that a case of slow collapse should account for the specific collapse. The rapid collapse, by the same logic, should account for some kind of a nonspecific (polymer-like) contraction, achieved purely by transfer of the disordered chain from good-solvent to poor-solvent conditions (Agashe et al. 1995; Sosnick et al. 1997; Qi et al. 1998). Such a model, like the previously mentioned de Gennes-model, seemed plausible, but then one notices that the experimental data for even the fastest-collapsing proteins also show some evidence for concurrent structure formation (Nolting et al. 1997; Sauder and Roder 1998; Hagen and Eaton 2000; Akiyama et al. 2002)! Observation became even more intriguing with the Klimov–Thirumalai assertion (Klimov and Thirumalai 1996), while talking about random heteropolymer model; that, "the kinetic accessibility of the native states in proteins may be encoded in the primary sequence". Such a finding, quite logically, hinted at the question that whether the existence of a stable native structure for a primary structure is necessary and sufficient to dictate the dynamics of the initial collapse? A subsequent study (Qiu et al. 2003) did probe this question in sufficient details to find that equilibrium stability of the native state does not control the collapse dynamics of primary structure.

Thus we can assert that a reasonable degree of understanding about protein folding has been obtained from studying it from the framework used for polymer collapse studies. But even after that, till to date, the bulk of unanswered questions embossing this paradigm tend to overshadow the number of answered questions. A detailed discussion on polymer collapse versus protein folding is out of the scope of present book. For fantastic treatment of this topic the interested readers can refer to (Dill 1990; Mirny and Shakhnovich 2001; Dill 1999; Shakhnovich 2006); the last two being of exceptional significance.

Getting back to fractal dimension-based investigation of proteins; we will now explore the FD approaches that attempt to study proteins without involving the backbone information. In these methods, proteins are considered more like clouds of residues (or clouds of atoms). By identifying the self-similarity in the distribution of these clouds of residues or atoms, the fractal dimension in the distribution of the relevant entity is calculated.

2.2.2 Approach with Residue (and Atom) Distribution

This approach of FD calculation disregards the information about connectivity profile of main-chain atoms and concentrates solely on the distribution of amino acid residues and atoms within the folded protein structures. Whether these residues and atoms are connected or not is not considered in this approach. This procedure (commonly called the 'mass fractal' calculation) studies the mass distributions within concentric spheres using, for example, the mass scaling relationship: $M \propto R^{FD}$ (where M is the total mass of all the protein atoms and R is the characteristic length scale). In this approach, from studying the linear (scale invariant) portion of the slope of $\log M$ versus $\log R$ plot, one can determine the FD. Theoretically, one can accomplish this by counting the number of atoms within concentric spheres, taking any atom of the protein space as center. However, for most of the practical applications, choice of center of mass of the protein for the role of aforementioned center makes more sense. Though the generic name for calculation of this genre of FD is mass fractals, one can calculate other pertinent properties associated with protein atoms with this methodology too. More magnitude of mass-FD for any such property implies that, the property under consideration is more space filling in protein interior. The mass-FD magnitude varies as 2.00 < mass-FD < 3.00. Proteins have an intrinsic self-similarity with respect to their compactness (atom packing) (Enright and Leitner 2005; Banerji and Ghosh 2009b). Concept of mass-FD builds upon this fact, it simply quantifies the symmetry of scale-invariance associated with any protein property that is dependent upon atom packing. It can be extended to the level of residues too, but not to the realm of organization of residues in secondary structures (α-helices are not composed of smaller helices).

This approach of FD calculation is rather new. Although one finds its reference in Dewey's work (Dewey 1997), the first systematic attempt to describe proteins with this methodology could only be observed as recently as in 2005 (Enright and Leitner 2005). However, from this point onward, lot of studies (Reuveni et al. 2008; Banerji and Ghosh 2009b; Moret et al. 2009; Lee 2006; Figueiredo et al. 2008; Hong and Jinzhi 2009) have utilized and extended the scope of this powerful construct to successfully unearth many forms of latent symmetries in the paradigm of protein interior interactions (these are discussed later).

The mass-FD calculation, even though presents a completely different standpoint of interior fractal study, is related to the approaches discussed above. One can trace back the philosophy of calculation of mass-FD to principles of polymer physics too. Criterion of a "good" solvent demands the polymeric units to have preferential interaction with the solvent, rather than with each other. As a consequence, the polymer acquires an extended shape and is called "excluded volume polymer". The Flory constant v_F, in this case assumes the magnitude $3/5$. In an "ideal" solvent (the so-called θ-solvent), polymer interacts with the solvent with identical strength, as it interacts with itself. As a result, in an "ideal" solvent, polymer assumes a more compact shape than it does in a "good" solvent, with $v_F = 1/2$. Polymer collapse takes place in the presence of a "poor" solvent. In such a case polymer–polymer interactions become more favorable than polymer–solvent interactions. Hence the polymer forms a collapsed or a globule state with $v_F = 1/3$. It is easy to notice that this entire spectrum of possibilities can be elegantly described by: $R_g \sim N^{v_F}$ —where R_g denotes the ROG of the polymer chain (a measure of its compactness), N stands for the number of units and Flory constant v_F assumes different values depending upon the solvent conditions. But $N \sim M$ and $M \sim R^{FD}$. Hence, we end up with $v_F = 1/FD$. Thus, the excluded volume polymer will have FD $= 5/3$, for an ideal polymer the FD magnitude will assume that of the topological dimension, viz. FD $= 2$; whereas for a collapsed polymer, the FD will assume the magnitude of another topological dimension, viz.FD $= 3$; whereby the polymer will assume a homogeneous space-filling nature. Neither $v_F = 1/2$ nor $v_F = 1/3$ is pertinent in the context of a functioning protein; in fact, they merely serve as the trivial limiting cases. Thus, while the relationship (2 < mass − FD < 3) holds for proteins, one can effortlessly connect fledgling mass-FD studies with the established set of treatise of polymer physics too. We will later talk about mass fractals in details. For a comprehensive review of polymer studies with FD measures, one can refer to a (classic) decade-old review (Novikov and Kozlov 2000).

(A note of caution, the previous few lines might provide the readers with an impression that the concept of FD is synonymous with that of ROG, but it is not so. The reasons for their being different are discussed in detail, afterward. Since most of the polymer physics works refer to ROG as a measure of compactness (instead of mass-FD), to draw the connection, this discussion on ROG was included.)

2.2.3 Approach with Correlation Dimension Analysis

There are situations where one has a qualitative idea about the distribution of certain properties within the protein space, but existences of a possible symmetry in their dependencies are not known. To obtain an objective idea about the distribution of the dependencies, it is convenient to opt for the study of a different fractal dimension, the 'Correlation dimension'(CD). Informally, to determine the CD, the pair of points between which possible dependencies may exist are identified at first. Amongst these, the ones that lie within a sphere of radius r are counted, say this number is denoted by $C(r)$. Then, by assuming a scaling $C(r) \sim r^{CD}$, the Grassberger–Procaccia algorithm (Grassberger and Procaccia 1983) is implemented to examine the possible presence of a linear stretch in the slope of a log–log plot of $C(r)$ versus r. Although a previous work (Lee 2008) had characterized the roughness of protein surfaces with CD, application of the same for protein interior studies started only very recently (Tejera et al. 2009), where it was shown that CD magnitude for ribosomal RNAs could successfully explain their structural characteristics (local helix formation and long-range tertiary interaction forming three-dimensional structures).

A categorical implementation of CD can be attempted by suitably describing each of the N monomers by some typical atom (say the alpha carbon atom C^α or centroid of the residue). Describing these atoms by points, vectorial distance between two monomers i and j can be denoted as $|\vec{x}_i - \vec{x}_j|$. Then one can formally define CD as:

$$\text{Corr}(r) = \frac{2}{N(N-1)} \sum_{i<j}^{N} \theta\left(r - \left(|\vec{x}_i - \vec{x}_j|\right)\right)$$

where, $\theta(x)$ is the Heaviside step function and $N(N-1)$ is a renormalization term (presence of which implies that Corr(r) can be considered as the probability that any two residues are in contact at a cut-off distance r. The term $\left[2\sum_{i<j}^{N} \theta\left(r - \left(|\vec{x}_i - \vec{x}_j|\right)\right)\right]$ describes the symmetrical properties of the contact matrix. In fact, Corr(r) is an unbiased estimator of the correlation integral:

$$C(r) = \int d\mu(x) \int d\mu(y)\, \theta(r - (|x-y|))$$

Both Corr(r) and C(r), monotonically decrease to zero as $r \to 0$. If C(r) decreases with a power law, viz., $C(r) \sim r^D$; then D is called the correlation dimension of μ. Comprehensive theoretical discussion on this can be found elsewhere (Takens 1985). Finally, the dimension is defined by: $D = \lim\limits_{r \to 0} \frac{\log(\text{Corr}(r))}{\log(r)}$. Details of implementation of this algorithm along with thorough discussion of several aspects of it can be found in (Lee 2008) and (Tejera et al. 2009).

The discussion above simply tells us how to calculate CD, but what exactly does correlation dimension mean? Since the correlation dimension attempts to quantify the self-similarity in the distribution of dependencies, it is important to understand how these dependencies are identified, and also how these dependencies are contingent upon the distribution of atoms or residues or other protein structural entities. An atom or a residue can be described by a point, depending upon the resolution at which one is interested to quantify protein symmetry. Every atom of protein is endowed with certain properties; say the mass of its, the 'residue specific atomic hydrophobicity' of its, the electronegativity of its, etc. Similarly, every residue is endowed with a set of property of its too; residue hydrophobicity, residue polarizability, the Chou–Fasman propensity of the residue, etc. Now, suppose one wants to calculate the correlation dimension between the positively charged amino acids in a protein. Since the person concerned is interested to quantify the extent of self-similarity in the distribution of positively charged amino acids, as the first step, he (she at any rate) will have to identify the coordinates of each of Arg, His, Lys residues in that protein. Drawing a sphere of radius r and counting the number of pairs of any two positively charged amino acids simply accounts for counting the possible pairs between any two of the three positively charged amino acids within a distance threshold. Denoting the number of pairs thus obtained by $C(r)$ and assuming the scaling $C(r) \sim r^{\text{CD}}$, the Grassberger–Procaccia algorithm can be implemented to examine the possible presence of a linear stretch in the slope of a log–log plot of $C(r)$ versus r. But the process of identifying the number of pairs of positively charged residues within a distance threshold of r Å is the same as identifying the extent of dependency (or correlation) in the spatial disposition of positively charged amino acids. Thus, although the Grassberger–Procaccia algorithm requires merely the coordinate information about pertinent entities, in the context of proteins, in effect, the correlation dimension is quantifying the extent of symmetry of self-similarity in the spatial distribution of any property under consideration. One can of course extend the scope of applicability of this powerful algorithm by attempting to innovatively quantify the self-similarity in the distribution of other protein structural correlations (or alternatively, protein structural dependencies). For example, a recent work has calculated the correlation dimension in the distribution of peptide dipole moments (Banerji and Ghosh 2011). However, one may point out that there are other (somewhat) similar constructs, say residue contact order (CO), residue contact map etc., which attempt to unearth similar array of information. Since, like inverse problems, all of them attempt to bring to light certain deeper truths about the nature of protein folding and about the nature of structural constraints found in native state of a protein, by merely studying the short-range and long-range contacts between the residues. To what extent the correlation dimension is similar to them and by how much do they differ?—These are important questions. The next section is entirely devoted to discuss these issues.

2.2.3.1   Residue Contact Order, Contact Map, and Correlation Dimension

Before we start our discussion on constructs that bank on residue contact profiles, it assumes significance to talk a little bit about the 'framework model' of protein folding and about importance of local interactions in protein folding and protein stability. Such an approach will help us to relate characteristics of local interactions to residue contact profiles, and subsequently to protein folding. To appreciate the power of the construct correlation dimension, it is imperative (if not obligatory) to gather a thorough idea about distribution of local interactions and contact profiles in proteins.

Starting with the Ptitsyn–Rashin model (1975), the 'framework model' (Kim and Baldwin 1982; Ptitsyn 1987; Kim and Baldwin 1990) advocated that secondary structures form first during the process of protein folding. Such formations are followed by docking of the pre-formed secondary structural units to yield the native, folded protein. Readers may note that such a view of protein folding is starkly opposite to the other view, that of the 'hydrophobic collapse model' (Schellman 1955; Kauzmann 1959; Tanford 1962; Baldwin 1989), where hydrophobic collapse drives compaction of the protein so that folding can take place in a confined volume, thereby narrowing the conformational search to the native state. The framework model was supported by findings obtained from various studies mostly performed during 1980s on small, relatively stable, and helical peptides (Brown and Klee 1971; Bierzynski et al. 1982; Shoemaker et al. 1985). Assertion of "secondary structures first" from framework model was significant, because previously it was assumed that secondary structural segments were not sufficiently stable to form in the absence of tertiary contacts (Epand and Scheraga 1968). Subsequently, however, numerous incomplete aspects of framework model were identified, but we won't get into that discussion now. What is of interest for the present treatise is to note that, the idea of significance of local contact, and therefore of local interaction, which was implicit in the framework model of protein folding.

Precisely how important are the local contacts and local interactions? Before we attempt to answer this question from the (traditional) paradigm of folded proteins, let us note some other interesting findings. At least two reports (Dobson 1992; Shortle 1996) in the mid-1990s revealed that the denatured state is rarely a random, unstructured coil that our textbooks tend to make us think, where the side-chain interactions assume triviality and all amino acids behave (almost) independently. Far from it. It is been found that very harsh conditions are necessary to obtain such a classical "disordered" state. Instead, these works suggested, that proteins generally adopt residual structure, which may well be in the form of fluctuating secondary structures and dynamic side-chain interactions, particularly hydrophobic clusters (Dobson 1992; Shortle 1996). Furthermore, this residual structure can be native-like or non-native. Residual structure can be just a product of the *tendency* of the chain to limit its solvent accessible surface area, or the structure can be beneficial in folding by helping the protein to reach the native state. Since, in the absence of well-defined secondary structures, it is the local contact profiles and local interactions which define the template of the residual structures, the importance of local contact and local interactions could hardly be trivial, even while considering the denatured state.

Indeed, the nontriviality of local interactions could be established categorically in the paradigm of folded proteins too. Various studies on protein folding kinetics have demonstrated that some native-like secondary structure elements form early and rapidly (Wright et al. 1988). Such a finding implies unambiguously that some regions of polypeptide chain can adopt their native-like structure, even in the absence of interactions with the other chain(s) or other parts of the same chain. Such local structures can therefore be expected to have a crucial role in initiating of protein folding (Dill 1990; Baldwin and Rose 1999). Furthermore, they may assume importance in stabilization of the transition state structure also, which corresponds to the rate-limiting step leading to the folded state of native protein (Nölting and Andret 2000; Guerois and Serrano 2001).

Which are the secondary structures that form most readily?—Various studies have established that some short stretches of amino acids can form native-like α-helices and β-hairpins, independently and quickly, either in aqueous solution or in the presence of an organic co-solvent (Dyson et al. 1988; Blanco et al. 1993; Blanco et al. 1994). Formation of an α-helix occurs over a nanosecond time scale; for example, ~200 ns is enough to fold an α-helix of 20–30 residues (Williams et al. 1996; Gilmanshin et al. 1997; Eaton et al. 1998). Such a time scale agrees largely to the time scale of elongation (or shortening) of an α-helix by one residue, which is in the range of 1–10 ns per residue according to the ultrasonic measurements held with both long (Zana 1975) and short (Eaton et al. 1998) synthetic polypeptide). Time taken for β-hairpin construction is about 30 times more than that required to construct an α-helix (Munoz et al. 1997). The difference between these time scales can be attributed to the fact that in an α-helix, each new hydrogen bond fixes one new residue; whereas in β-hairpin, each new hydrogen bond fixes two new residues. Hence, to attain stability, a β-hairpin should necessarily have very specific complementarity of sequences of its β-strands, which dictates only one point of the chain turn; while the first turn of an α-helix can be anywhere in the chain. To gather more insights on this topic, an interested reader can refer to (Fukada and Maeda 1990) and (Finkelstein 1991). Furthermore, to obtain a comprehensive idea about how the inter-residue interactions (short, medium, and long range) play an important role in the folding and stability of globular proteins, one may refer to (Miyazawa and Jernigan 1999). Suffice to say that all of the aforementioned results unmistakably demonstrate that, importance of local interactions in protein folding and protein structure paradigm was known to biophysicists for a long time. Problem was in quantifying the extent of local interactions, so that they can be directly related to the paradigm of protein structures. In this context, we introduce ourselves to the concept of residue 'CO'.

By denoting the average distance between interacting residues in the native state as 'CO', in a landmark paper in 1998, David Baker and his colleagues (Plaxco et al. 1998) proposed it could be considered as a general descriptor of protein topology to correlate topology with the folding rate. Defined with the simplest of the schemes

$$CO = \frac{1}{N.L} \sum_{i,j} C_{i,j} |i - j|$$

where N is the total number of residues, L is the length of the protein, and $C_{i,j} = 1$ if residues i and j are in contact, $C_{i,j} = 0$ otherwise, the parameter CO could connect local interactions to secondary structures, and then to protein structure and protein folding—all by a single thread. (The contact map $Mc_{i,j}$ among the residues could accordingly be constructed by defining,

$$Mc_{i,j} = \begin{cases} 1 & \text{if } d_{i,j} \leq r \\ 0 & \text{if } d_{i,j} > r, \text{ or if } i = j \end{cases}$$

where r stands for the distance threshold and $d_{i,j}$ denotes the spatial distance between ith and jth residues.) Owing to the fact that the α-helical proteins have large number of local contacts, they have low CO. In contrast, proteins with preponderance of β-sheets, as also the α/β proteins, owing to their numerous nonlocal contacts, have high CO. More significantly, since the α-helical proteins have been shown to fold more rapidly (discussed previously) than the proteins with preponderance of β-sheets, or the α/β proteins—one could establish a negative correlation between folding rate and CO of residues. Formalization of this idea came when another work (Fersht 2000) proposed an extended nucleus mechanism of protein folding that related CO, chain topology and stability, and protein folding rates. (Alongside these, to obtain a comprehensive idea about how the inter-residue interactions (short, medium, and long range) play an important role in the folding and stability of globular proteins, the reader may refer to (Miyazawa and Jernigan 1999).) Thus, in one leap, the chasm between microscopic unit of the protein and its macroscopic self could be bridged. Applicability of the new concept was put to test (almost) immediately and two works reported straightaway that for quite a few proteins that demonstrate two-state folding, the CO correlates well with log of their respective folding rates in water (Jackson 1998; Munoz and Eaton 1999).

However, before long it has been found out that the 'absolute CO' (derived by simply dropping the normalization factor from the denominator in the aforementioned equation of 'CO'), instead of the originally proposed definition of CO, predicts the folding rates better. At about the same time two other measures (Mirny and Shakhnovich 2001; Gromiha and Selvaraj 2001) demonstrated that local interactions could be defined differently and then quantified efficiently by other constructs too. The first approach (Mirny and Shakhnovich 2001) attempted to quantify the fraction of local and nonlocal contacts by quantifying CO as:

$$CO = 1/L\left[f.L_{local} - (1-f).L_{Distant}\right],$$

where $f = N_{local}/N$, denoted a fraction of local orders where the condition $|Res_i - Res_j| \leq 4$ is satisfied; L_{local} denoted the average separation between residues forming local contact; and $L_{Distant}$ denoted the average separation between residues forming distant (viz. $|Res_i - Res_j| > 4$) contact. The second approach (Gromiha and Selvaraj 2001) attempted to quantify the long-range orders (LRO) between residues by adopting the scheme:

LRO $= \sum n_{ij}/N$, defining number of residues n_{ij} as $n_{ij} = 1$, if $|i - j| > 12$, 0 otherwise; and denoting the total number of residues in a protein by N. Since then, various permutations and combinations have been tried out with the aforementioned parameters in search of a robust construct that provides insights about overall structure and nature of folding by scanning through the sequence.

However, to obtain a comprehensive idea about importance of a pair of any two arbitrarily chosen residues, one needs to have something more than a geometric parameter; it can be a distribution function describing the global disposition profile of any structural entity, or it can be a measure that attempts to quantify the global symmetry of structural organization that connects the local geometric measures to (possibly) symmetric disposition of its global profile. This is where the biophysical importance of correlation dimensions can be realized. But the idea of correlation dimension does not stand isolated, one can indeed derive the expression of correlation dimension from the set of aforementioned (elementary) ideas.

One can attempt to connect the aforementioned ideas by a single thread to see how residue–residue contact studies can be extended to construct a framework that can unearth the symmetries in the prevailing dependencies among the various biophysical properties as manifested by the spatial arrangement of residues. As it has been talked about beforehand, the CO is simply defined as: $CO = \frac{1}{N.L} \sum_{i,j} C_{i,j} |i - j|$ (symbols retaining their meaning provided beforehand, the exact value of distance threshold can vary from case to case, thus we refrain from 6 or 8 Å debate). A modification of this scheme was proposed in the form of relative contact order (RCO), defined as: RCO $= \frac{\frac{1}{N_c} \sum_{i<j}(j-i)}{N}$ (where N_c is the total number of contacts between amino acids in a protein, N is the total number of amino acids and the sum is taken over all (properly defined) contacts between amino acids.). The contact map $Mc_{i,j}$ is the visual way to ascertain contact profile among the residues. The residue contact map be constructed by defining:

$$Mc_{i,j} = \begin{cases} 1 & \text{if } d_{i,j} \leq r \\ 0 & \text{if } d_{i,j} > r, \text{ or if } i = j \end{cases}$$ Since each amino acid embodies a distinct type

and extent of biophysical property (properties being size, hydrophobicity, electronegativity, etc.), one can obtain an idea about disposition of certain biophysical property by suitably constructing the contact map. This can easily be achieved by including certain residues that are relevant for the calculation, and not including certain other residues, on the ground that the later are not pertinent in the context of the biophysical property under consideration. Furthermore, one can even, with suitable modifications of expressions of CO and RCO, calculate the CO and RCO, with respect to the agents (viz. the amino acids) that embody the particular property under investigation. Hence, the aforementioned tools can present us meaningful information about residue arrangement, the nature of spatial dependency amongst the various properties. However, such a study will be residue centric and will fail to reveal the presence of any symmetry in the distribution of the property under consideration, at the global scale. To construct such a scheme one needs to connect the information of the location of a set of residues from local arrangement, to the arrangement prevailing at the biological functional state (found from biological unit) of the protein. This can be achieved by changing the focus from distances between residues paradigm to residues across the protein paradigm, which will be spatially isotropic and will potentially include all the residues but will also have the filtering option to consider only the residues that are relevant.

To achieve this, one can start with the contact matrix and then calculate the number of residues at particular distance thresholds from other residues. Since the nature of residue disposition is assumed to be isotropic, such an operation can be performed. Whereby, upon performing this operation with all the residues, we obtain:

$NC\,(r) = \sum_j^N Mc_{i,j} = \sum_j^N \theta\,(r - \|d_{i,j}\|)$, where θ is Heaviside function and N is the total number of residues. The total number of relevant contacts can then be calculated as:

$$NC\,(r) = \sum_i^N Nc(r)_i = \sum_i^N \sum_{j\neq i}^N Mc_{i,j} = 2\sum_{i<j}^N \theta\,(r - \|d_{i,j}\|).$$

Then, defining the correlation function as:$C\,(r) = \frac{2}{N(N-1)} \sum_{i<j}^N \theta\,(r - \|d_{i,j}\|)$, that is, equivalently:

$C\,(r) = \frac{1}{N(N-1)} NC(r)$ —one arrives at the expression of correlation dimension. Alternatively, this expression can be thought of as:

$C\,(r) = \frac{1}{C_2^N} \sum_{i<j}^N \theta\,(r - \|d_{i,j}\|)$, where the term C_2^N $\left(C_2^N = \dfrac{N\,(N-1)}{2}\right)$ denotes the combination of N amino acids that satisfies the distance threshold, taking any two at a time. Such an expression, derived from assuming equiprobable spatial distribution of amino acids, provides an ideal platform to study underlying symmetries existing between biophysical properties, where the biophysical properties themselves are embodied by amino acids. The CD, quantifying such symmetry, can be calculated from plotting $C(r)$ versus r—in the log–log plot.

One can connect the MFD with CD too. Starting with residues at any particular threshold from any arbitrarily chosen ith residue we arrive at:

$NC\,(r) = \sum_j^N Mc_{i,j} = \sum_j^N \theta\,(r - \|d_{i,j}\|)$. Now, looking at the problem from the perspective of mass distribution we arrive at :

$$\langle M(r)\rangle = \frac{1}{N}\sum_i^N \sum_{j\neq i}^N \theta\,(r - \|d_{i,j}\|) = \frac{1}{N}\sum_i^N \sum_{j\neq i}^N Mc_{i,j} = \frac{1}{N}NC\,(r) = (N-1)\,C(r)$$

Since we know the mass of a protein varies as:

$\langle M(r) \rangle = r^{\text{MFD}}$, we find: $C(r) \sim r^{\text{CD}} \sim r^{\text{MFD}}$

However, such a derivation will fail to provide the physical implications of CD, the implication with which symmetry amongst residues exhibiting certain properties can be studied. Hence, the relationship between MFD and CD should not be expected to reveal much valuable information.

2.2.4 Approaches with RGT

As has already been mentioned, proteins are large macromolecules, and protein structures undergo large-scale fluctuations. In fact, it has been reported in many cases that such large-amplitude movements are pivotally important for certain proteins to ensure proper functioning (Steinbach et al. 1991; Rasmussen et al. 1992). Since these fluctuations occur over all length scales, determination of a single characteristic scale parameter becomes extremely difficult (if not impossible). As a result, correlation functions between various biophysical properties within proteins typically demonstrate nonanalytical behavior. More often than not, these correlation functions tend to follow fractional power law scaling [for an explanation, please refer to (Dewey 1997)] and depend on the dimensionality of the protein (under a given context), rather than on its microscopic characteristics.

Renormalization methods are widely used methods in theoretical physics. One can loosely say that renormalization theory models the flow of the parameters in the parameter space describing a dynamical system. In condensed matter physics, critical exponents derived by the renormalization method have been shown (empirically) to have strong "universality"(Stanley 1999; Benney et al. 1992), which requires that the systems have same dimension and that the numbers of states have the same critical exponent. Thanks to the universality thus identified, the critical exponents of the systems can be considered as reliable measures for classifying dynamical systems; the classes, thus defined, are accordingly called universality classes.

Although the RGT was originally applied to describe critical phenomena (Wilson 1975), the study of fractals and RGT is naturally linked because as in the case of fractal studies, the concept of scale invariance plays a major role in RGT. To talk about it briefly, the absence of an internal scale is one of the fundamental features of fractals. This can be rephrased by asserting that in fractals a great many number of scales coexist in a self-similar manner, which in turn reminds us of a comparable situation observed in thermodynamic systems at the critical point. Divergence of correlation length at the critical point can be mathematically handled by assuming a composition of subsystems with some special coupling constant (Wilson 1979). In such a case, each of the subsystems can be assumed to be built up of sub-subsystems with another coupling constant, and so on—which can reliably be called the self-similarity of protein interior organization. While FD-based studies attempt to quantify this self-similarity, the central idea of the RGT is that a change in the scale should lead to the same behavior apart from a renormalization of the coupling constant.

Two principal transformations are studied in the RGT paradigm, namely, the coarse graining and the scaling (Goldenfeld 1992). During the analysis of large-scale fluctuations of a protein, studies of its microscale perturbations may not always assume a great importance. Hence, in order to tackle the huge complexity associated with the fluctuations of proteins, researchers have come up with the efficient strategy of removing the length of the shortest interactions before the system is redefined by rescaling the length scale (Dewey 1997). Since proteins exhibit self-similarity in their organization, this self-similarity can be captured with RGT and used to probe the various facets of (time variant and context dependent) interior dependencies. A categorical example of this can be found from the analysis of protein folding, which has been proved to be comparable to a phase transition (Goldstein et al. 1992a, b). These problems are characterized by a multiplicity of scale lengths [one observes a large spectrum of correlations that correspond with the critical point, from short-range (among residues close by) to long range (amongst residues situated far apart)]. This discussion of the scope of RGT implies that it can be suitable for a rigorous and objective description of such phenomena. Indeed, taking a cue from polymer physics studies (Family 1982), RGT has also been applied in the realm of protein biophysical organizational studies (Pierri et al. 2008; Bohm 1991; Li et al. 1990a, b; Nonnenmacher 1989; and references therein).

The (possible) relevance of direct applications of renormalization methods in protein structure studies can be estimated from the fact that multi-scale techniques have emerged as potent tools to combine the efficiency of coarse-grain simulations with the detail of all-atom simulations for the characterization of various real-life systems, such as proteins. For example, such an idea has been applied to a system of small molecules where some parts of its space (that is, parts of the molecular space with more relevance to the problem undertaken) are described with the all-atom representation; while the rest of the space, which influences the properties of more important part but is not critically important by itself, is described through coarse-grain representation (Praprotnik et al. 2005). Similar type of multiple resolution simulations have also been employed to investigate membrane-bound ion channels by coarse graining the lipid and water molecules while using an all-atom representation for the polypeptide ion channel (Shi et al. 2006). In the realm of protein structure, a similar idea has been used to represent parts of the protein, where the active site is described in all-atom detail and rest of the protein was described with a coarse-grain model (Neri et al. 2005). Interestingly, one finds that such line of thought is not exactly new. In search of an appropriate construct to model the scaling, a 1999 paper (Fan et al. 1999) attempted to change the whole system resolution during the same simulation (although the word "scaling" was not used in the whole paper). Starting with a simplified protein model, which differed from the then-contemporary descriptions of simplified proteins significantly, the work evaluated the folding free energy of the corresponding all-atom model.

In early 1990s, Bohr and Wolynes (Bohr and Wolynes 1992) employed it for a protein folding study with contact energies, while Chan and Dill (Chan and Dill 1991) described its possible effectiveness for protein stability studies. Subsequently, another work (Coveney and Fowler 2005) has used it for 'systematic coarse-graining' of a system of coupled nonlinear ordinary differential equations (see also relevant references therein) while studying protein interaction profiles. However, even after all these applications, the use of RGT is yet to become a common practice while attempting the scaling and coarse-graining operations in protein paradigm; as expressed in a recent work: "rigorous mathematical procedures, such as RGT, have yet to be applied to the general definition of coarse-grain models" (Heath et al. 2007), while studying proteins.

Since FD and RGT-based studies are closely related, there is an overlap between these approaches, and indeed many FD-based studies have explored their interface (Song et al. 2006; Kitsak et al. 2007; Rozenfeld et al. 2010) with serious intent. It should not be forgotten, however, that these studies have been performed on a protein network paradigm and not on protein structures. Although there are two fantastic works (Dewey 1997; Freed 1987) in which the approaches and results of FD and RGT (in the context of polymer and proteins) are compared and contrasted in admirable details. In the sphere of protein interior studies, one merely finds a small number of FD studies that tangentially mention RGT (Havlin and Ben-Avraham 1982b; Wagner et al. 1985; Colvin and Stapleton 1985). In order to truly assess the utility of renormalizations methods to study protein structural and folding properties, one needs to estimate the feasibility of consistently adding all-atom detail to coarse-grain protein models. Especially, one needs to investigate whether, the shift from coarse-grain to all-atom description distorts the thermodynamic properties of the corresponding ensembles of structures—or not. Only after these issues are resolved, can we expect to find works that employ renormalization methods to unearth protein-scaling symmetries.

2.2.5 Approach with the Spectral Dimension

Here we talk about the dialectics of protein structures! Had he known about it, Frederick Engels would have surely incorporated this idea in his magnum opus 'Anti-Dühring'; (fortunately) we knew not much of physics of protein structure back then. Jokes apart, are there fundamental contradiction that the protein native structures suffer from? As it turns out, there is (at least) one, which can be stated in two equivalent and dependent ways; one in terms of dynamics, the other, in terms of potential energy.

Let us start with the potential energy centric formulation. If a protein is in its native state, the compactly folded structure of its can be expected to assume a single well-defined equilibrium structure. In such a case, the overall shape of the potential energy surface must be that of a single well. But it is well-known that the stable equilibrium configurations of proteins actually consist of multiple conformational substates (Elber and Karplus 1987; Frauenfelder et al. 1988; Kitao et al. 1998). The problem, then, is to ascertain the extent to which the "fine structural" aspects of the potential energy surface can be regarded as a modulation of a smooth single-well potential which defines the protein's general folding scaffold. The driving force is that, if indeed it is found out to be so, then (in the best of the two-worlds description) motion in such a system can be described qualitatively as vibrational motion within a local potential energy minimum on a short-time scale, and as diffusion on a smooth effective potential energy surface on a long-time scale.

The same problem can be looked at from the viewpoint of protein dynamics also. On the one hand, proteins attempt to maintain the structure of their native fold thermally stable. On the other hand, such native fold template needs to accommodate large amplitude conformational changes that allow appropriate functioning of the protein (Karplus and McCammon 1983; Bahar et al. 1998; Henzler-Wildman et al. 2007; de Leeuw et al. 2009). The extent of such conformational changes can be assessed by noticing that some proteins have been reported to exist in two or more number of equilibrium states relevant to their function (Damaschun et al. 1999; Frauenfelder and McMahon 1998). The two properties are not independent of each other; instead, it has been found that fluctuations in densely packed regions manipulate the motion of flexible parts of proteins. One may argue that although interesting, these are rather predictable findings. What, however, could not be labeled as predictable was the discovery that the coexistence of the aforementioned contradiction could not be explained by considering proteins as compact objects, since the later are necessarily characterized by small amplitude vibrations (Burioni et al. 2004; de Leeuw et al. 2009). To appreciate why these facts are extraordinary we need to look at the whole perspective that connects residual behavior with folded topology.

Although residual and atomic arrangements within proteins have distinct geometrical nature, such arrangements are far from being static. A sizeable accumulation of experimental data collected from NMR and neutron spectroscopy has revealed that protein native states are dynamic structures wherein amino acids (and the atoms therein) constantly move around their equilibrium positions. Furthermore, it has been established that this motion acts as a crucial determinant toward ensuring protein function (Frauenfelder et al. 1979; Frauenfelder and McMahon 1998). Theoretical and computational studies on fluctuations and collective motions of proteins are (usually) conducted with either molecular dynamics (MD) simulations or normal mode analysis (NMA) (Levitt et al. 1985; Bahar and Rader 2005) or essential dynamics (Amadei et al. 1993). Detailed discussion about them is not possible to undertake here, but we will briefly talk about NMA.

In NMA paradigm, proteins are modeled as elastic networks, where nodes denote the residues, linked by inter-residue potentials that stabilize the folded conformation. Residues, in turn, are assumed to undergo (Gaussian-distributed) fluctuations about their mean native positions. The springs connecting each node to all other neighboring nodes are assumed to be of equal strength, and only the atom pairs within a distance threshold are considered without making a distinction between different types of residues. Although highly simplified, using a single parameter harmonic potential (Cui 2006), the NMA successfully predicts the large amplitude motions of proteins in the native state (Bahar et al. 1997). Looking at it the other way round, one may assess the shear importance of interior fluctuations by the fact that relying mostly on geometry and mass distribution of the protein, the NMA could demonstrate that a single-parameter model can reproduce the complex vibrational properties of macromolecular systems. Of notable relevance to the present discussion is the finding that separating different components of normal modes, say, the collective (low-frequency) motions, the nature of a conformational change, for example due to the binding of a ligand, can be analyzed in depth (Delarue and Sanejouand 2002).

But how are all these related to fractal studies of proteins? Well, there are two ways to establish the relevance: first, from the perspective of fluctuation and dynamics, second from the perspective of energy flow. We will start with the fluctuation and dynamics part. Obtaining a thorough idea about the fluctuation dynamics of proteins near native state is necessary toward gaining insights about the molecular basis and mechanisms of their function. Studying the fluctuation dynamics becomes far easier with linear models such as NMA. Furthermore, to study the collective and correlated nature of residue fluctuations, the NMA presents itself as a reasonable first approximation. Moving on to the other aspect, viz. flow of energy studies, it was established that energy flow within a protein could be reliably described in terms of transfer among the collective oscillations, or normal modes, of the protein. Energy flow among the vibrational states of a protein (defined in terms of excitations of the vibrational modes) can be generated by anharmonic coupling between them, giving rise to energy flow pathways through the vibrational states. More importantly, it was found (way back in eighties) that low-frequency modes in proteins in the normal mode density, vary with frequency as expected for fractal objects (Elber and Karplus 1986; Elber 1989)! Although studies about the nature of energy equipartitioning in polymers are nothing new (Fermi et al. 1955; Uzer 1991), the importance of the last fact can only be assessed when we talk about flow of vibrational energy in proteins. It is here that we will be introduced to the third type of fractal dimension relevant in protein structure paradigm, the spectral dimension.

Gaining an understanding about energy flow in proteins is crucial requirement toward understanding protein stability and protein function. While one study attempted to address this problem by trying to identify evolutionarily conserved pathways of energetic connectivity across protein families (Lockless and Ranganathan 1999), another approach (Sharp and Skinner 2006) resorted to pump-probe MD, whereby exciting a selected set of atoms or residues with a set of oscillating forces the transmission of energy to other parts of the protein was probed using Fourier transform of the atomic motions. Many other works (Suel et al. 2003; Clarkson and Lee 2004; Fuentes et al. 2006—to name only a few) attempted to decipher the characteristics of energy flow in particular cases. Commonality amongst all these studies was that they attempted to decipher the connection between protein architecture and pathways for energy to reach a specific target. Such studies highlighted the role of allostery. However, although informative, such studies could not present a general framework of energy flow in proteins, which could consistently connect the rich dynamics of proteins with their energy landscapes. Such a general treatment of the problem could be obtained from an alternative view of energy flow among the collective oscillations of a protein (Moritsugu et al. 2000; Leitner 2001; Leitner 2002; Moritsugu et al. 2003; Fujisaki and Straub 2005). The latter view relates to vibrational energy flow in proteins. Vibrational energy flow has been identified as an important (if not fundamental) mechanism by which a protein maintains a near constant temperature while taking part chemical reactions in cells.

How to study the flow of vibrational energy within proteins? It was found out that energy transport on a percolation cluster (Alexander and Orbach 1982; Rammal and Toulouse 1983; Nakayama et al. 1994), in which energy flows readily between connected sites of the cluster and only slowly between weakly connected sites, reliably describes the energy flow in proteins. But since energy may be trapped in a percolation cluster, speaking statistically, the spread of energy assumes the form of anomalous subdiffusion in them. (In the case of normal diffusion, the mean squared displacement of particles varies linearly with time, viz. $\langle \Delta x^2(t) \rangle \sim t$. However, in some complex disordered media, diffusion process assumes an anomalous nature; whereby we observe $\langle \Delta x^2(t) \rangle \sim t^\alpha$. In the last equation, the scaling exponent $\alpha = 1$ characterizes anomalous diffusion. For $\alpha < 1$, the process is characterized as subdiffusion, and for $\alpha > 1$, it is called super diffusion. Differential equations of fractional order are well suited for describing fractal phenomena such as anomalous diffusion in complex disordered media. The constant memory and self-similar nature of these phenomena can be taken into account by using the kinetic equations with fixed fractional order.) Intriguingly, anomalous subdiffusion could be observed in simulations on proteins, as with a percolation cluster (Bouchaud and Georges 1990; Garcia et al. 1997); whereby, protein's geometry, its vibrational modes, energy flow through protein— could all be connected and it could be established that percolation cluster-based protein modeling could be reliable and beneficial. Amongst various other interesting properties, we will just mention that a general feature of vibrational energy flow among vibrational states of a molecule is its local nature (Logan and Wolynes 1990; Lehmann et al. 1994; Leitner and Wolynes 1996). Importance of the last observation can be assessed once the concept of spectral dimension is discussed.

In their pioneering study, Alexander and Orbach (Alexander and Orbach 1982) found that the mean square displacement (denoted by R^2) of a vibrational excitation on a fractal object varies as:

$$R^2 \propto t^\gamma, \text{where } \gamma = \frac{d}{D}$$

The denominator of γ, that is D, is (our old friend) mass fractal dimension (MFD); the numerator of γ, that is d, is the spectral dimension, whom we'll denote by SD from now. The last equation can therefore be written as:

$$R^2 \propto t^\gamma, \text{where } \gamma = \frac{\text{SD}}{\text{MFD}}$$

MFD characterizes the scaling of mass M with the backbone length L (that is: $M \sim L^{\mathrm{MFD}}$). The SD quantifies how the vibrational density of states $(g(\omega))$, varies with mode frequency (ω), viz. $g(\omega) \propto \omega^{\mathrm{SD}-1}$. The MFD quantifies how much of a space is filled up with the property under consideration; where the property under consideration can be mass, hydrophobicity, polarizability, chirality, etc.,—it does not necessarily have to be mass only. The SD quantifies the density of vibrational states. But then, how to count the number vibrational states with frequency less than (ω)?—That can be found from a parameter named 'cumulative density of states', $(G(\omega))$, which is defined as: $G(\omega) = \int_0^\omega g(\omega_{\mathrm{low}}) \, d\omega_{\mathrm{low}}$. While studying these low frequencies, one can resort to the scaling $(G(\omega)) \sim \omega^{\mathrm{SD}}$. For an infinite regular square lattice, the SD coincides with the regular dimension of 2; but for Sierpinski gasket it is smaller than 2 and is given by: $2\ln(3)/\ln(5) \approx 1.365$. For more discussions on it, the interested reader can straightaway look at the classic papers (Alexander and Orbach 1982) and (Rammal and Toulouse 1983).

The vibrational normal mode spectra, $(g(\omega))$, of proteins was first studied by Ben-Avraham for five proteins with sizes in the range of 39–375 residues. It was found that the data collapsed to produce a single curve, especially in the slow mode region (Ben-Avraham 1993). The density of states was found to increase linearly with the frequency in this region, implying a spectral dimension of magnitude SD = 2. This trend demonstrated a marked deviation from the Debye model of elastic solids, according to which the expected value should have been 3 (Kittel 2004). Subsequently, the anomalous spectral dimension of proteins was confirmed by inelastic neutron scattering experimental measurements, which yielded SD ≈ 1.4 for hen egg white lysozyme (Lushnikov et al. 2005). Very recently, in a work of far-reaching consequence to the study of protein structure physics, an equation of state relating the spectral dimension, fractal dimension, and the size of a protein has been proposed, based on the coexistence of stability and flexibility in folded proteins (Reuveni et al. 2008). The SD, therefore, can be considered as a reliable marker to study the low frequencies in vibrational dynamics of a fractal. For a more rigorous treatment of SD, the interested reader can refer to (Triebel 1997).

2.2.5.1 How to Calculate Spectral Dimension of a Protein?

This discussion is rooted in protein dynamics, for which we will have to revisit NMA of protein structures.

In a significant observation, in 1996, Tirion (1996) proposed the possibility of replacing the complicated empirical potentials in protein normal mode computations, with 'Hooke'-type pairwise interactions depending on a single parameter. Such an approach originated from the observation that low-frequency dynamics, which are mainly associated with protein-domain motion, are generally insensitive to the finer details of atomic interactions. Subsequently, many works (to name a significant few, Bahar et al. 1998; Jacobs et al. 2001; Micheletti et al. 2002; Banavar and Maritan 2003) have demonstrated the success of simple harmonic models in the study of the slow vibrational dynamics of large biological macro-molecules, and they have become a viable alternative to demanding and time-consuming all-atom NMA.

Going a step further, let us choose to describe a general elastic network of masses and harmonic springs. Physics tells us that such a network can be described by a set of normal modes and a corresponding set of eigenfrequencies. To recapitulate, a normal mode of an oscillating system is a particular type of motion in which all parts of the system move sinusoidally with the same frequency. The frequencies of the normal modes of a system are called its natural frequencies or eigenfrequencies. For such a system, the SD provides the information regarding density of low-frequency normal modes on a fractal. That is, there holds a relationship $g(\omega) \propto \omega^{SD-1}$ between density of modes ($g(\omega)$) and mode frequency (ω), for low frequencies.

At this point, we choose to talk about the 'Gaussian network model' (GNM) (Bahar et al. 1997). It is especially noteworthy in the present context, not merely because it yields results in agreement with X-ray spectroscopy experiments, but also because GNM has emerged as a potent system to study the spectral dimension. The GNM explores the role and contribution of purely topological constraints, defined by the 3D structure, on the collective dynamics of proteins, where the proteins are modeled as elastic networks whose nodes correspond to the positions of the C^α atoms in the native structure, and the interactions among nodes are described by (homogeneous) harmonic springs. These springs determine the degrees of freedom of the elastic network, viz. that of a protein structure, modeled as elastic network in present case. Furthermore, these springs govern the modes of vibrations of the given structure. One may wonder, what is the necessity for such kind of a model, when the crystal structure information for a protein is known?— Well, X-ray crystallography provides accurate structural information by allowing the determination of the average position of atoms and the amplitudes of their displacements from these average positions. But this analysis tells little about the ways the residues and the atoms move. Studying proteins with any form of elastic network model (like the GNM), assumes significance because such an investigation throws light on protein dynamics. Ironically, to verify the accuracy of such a model, the crystallographic information is sought. That is because, crystallographic structure determination includes obtaining information about thermal and other fluctuations of the atoms in a crystal. Typically, each atom can be assigned a Debye–Waller temperature factor or B-factor, with the latter proportional to the mean square amplitude of the fluctuations. Although these factors have some limitations (Kuriyan et al. 1986), they represent a very reliable experimental source of information on the dynamics of proteins. We'll find out shortly how this piece of information could be used for verification of GNM.

In GNM, an interaction between two nodes is considered to exist if the nodes are separated by a distance less than the pre-defined interaction threshold distance [usually, 6 or 7 Å (Burioni et al. 2004)]. The GNM Hamiltonian, thus constructed, is based upon two important assumptions; viz. residual fluctuations are isotropic in nature, and they are Gaussian in nature. Having said that, since the mean square

fluctuation for every residue is measurable with GNM, and since such a parameter
can be directly associated to B-factor of a residue in the studies of X-ray crystallography, since the GNM-predicted fluctuation profile showed excellent match
with crystallographic finding,—one can safely infer that those assumptions do not
differ much from biophysical reality. Indeed, studies examining the isotropic case
(Bahar et al. 1997) and anisotropic case (Kondrashov et al. 2007; Micheletti et al.
2002), revealed good match with theoretical predictions at room temperature. The
Hamiltonian for such a system is given by:

$$H_{GNM} = \sum_i \frac{\left(\vec{p_i}\right)^2}{2m} + \frac{k}{2} \sum_{i,j(j>i)} M_{i,j} \left(\Delta \vec{R}_i - \Delta \vec{R}_j\right)^2$$

Here, in the first term of the expression, 'p' represents the momentum and 'm'
represents the mass; whereby, being summed over all the residues, the first term
describes kinetic energy of the system. In the second term, 'k' represents the force
constant (assumed to be homogeneous), $\vec{R}_i$ and $\Delta \vec{R}_i$ are the instantaneous position and displacement with respect to equilibrium position $\vec{R}_i^0$ of the ith C^α atom.
Here we note that $\vec{R}_i$ and $\vec{R}_i^0$ are external coordinates, whereas $\vec{R}_{i,j}^0$, viz. the
separation vector between ith and jth residue in equilibrium—is an internal coordinate. (The $\Delta \vec{R}_i$ and $\Delta \vec{R}_j$ can be alternatively considered as instantaneous fluctuation vectors.) The 'M' is an old-friend of ours, the network connectivity matrix,
whom we had met during our discussion on correlation dimension. Just like the
case encountered then, we define $M_{i,j} = 1$, only when both the conditions $i \neq j$
and the distance $\left|\vec{R}_i - \vec{R}_j\right|$ between two C^α atoms in native state is less than some
pre-defined distance threshold—are met; otherwise $M_{i,j} = 0$. Since no distinction
is made between different types of amino acids, a generic spring constant k could
be adopted for modeling the interaction between all pairs of residues sufficiently
close. The significant aspect of this Hamiltonian is that not only the changes in
inter-residue distances, but also any change in the direction of inter-residue vector—will have a penalty to pay in GNM potential.

Thus the GNM, in short, describes protein mobility in terms of the atoms'
local packing density. Another peculiar feature of GNM is how the issue of
degrees of freedom is addressed here. That is, a protein is viewed here as a col-
lection of N sites, one for each residue, resulting in an ensemble of N—1 independent modes, instead of the $3N$—6 modes that would have been expected
from a 3D description. To obtain an idea about various other physical and computational aspects about low-frequency modes, NMA and GNM, the interested
readers can refer to (Haliloglu et al. 1997), (Hinsen 1999), and (Atilgan et al.
2001). Also, to obtain an idea about how the two elastic network models (NMA
and GNM both) of proteins have allowed us to observe that a few low-frequency
normal modes can present accurate information about the deep patterns in the
large amplitude motions of proteins upon ligand binding, demonstrating thereby
the robust character of residual collective motions, one can refer to (Tama and
Sanejouand 2001; Delarue and Sanejouand 2002; Krebs et al. 2002; Lu and
Ma 2005; Tama and Brooks 2006; Nicolay and Sanejouand 2006). Finally, to

acquire a thorough idea about the theoretical background behind GNM (and elastic networks in general), the interested reader can refer to (Eichinger 1972) and (Kloczkowski and Mark 1989).

A short note on (an)isotropy in residue fluctuation in GNM assumes importance here. Residue fluctuations are implicitly assumed to be isotropic in the GNM; and such an assumption did not produce results that differed by much from experimental data. However, in reality, it is known that residue fluctuations are in general anisotropic in nature (Kuriyan et al. 1986; Ichiye and Karplus 1987). Although an extension of the GNM, called the anisotropic network model (ANM), exists (Doruker et al. 2000; Atilgan et. al 2001), a satisfactory explanation of how could, then, the experimental results matched the results predicted through isotropic fluctuation assumption—is not easy to find. But one can (somewhat crudely) attribute the emergence of perceived isotropy to the fact that at systemic level, the minute effects of particular anisotropies won't probably be affecting the description appreciably.

Going back to the GNM Hamiltonian, let us zoom in on the interaction term, viz. $\frac{k}{2} \sum_{i,j(j>i)} M_{i,j} \left(\Delta \vec{R}_i - \Delta \vec{R}_j \right)^2$. Furthermore, let us concentrate only on the system-wide residual fluctuation part, viz. on $\sum_{i,j(j>i)} M_{i,j} \left(\Delta \vec{R}_i - \Delta \vec{R}_j \right)^2$. —This term expands as:

$$\sum_{i,j(j>i)} M_{i,j} \left(\Delta \vec{R}_i - \Delta \vec{R}_j \right)^2 = \sum_{i,j(j>i)} M_{i,j} \left(\Delta R_i^2 - 2\Delta \vec{R}_i \Delta \vec{R}_j + \Delta R_j^2 \right)$$

—which, sadly, isn't exactly very insightful. However, this term can now be subjected to threshold filtering, so that an effective contact matrix, say, 'C' can be constructed. To achieve this, we define:

$$C_{ij} = -1, \quad \text{if } i \neq j \text{ and } \vec{R}^0_{i,j} \leq \text{Threshold},$$

$$C_{ij} = 0, \quad \text{if } i \neq j \text{ and } \vec{R}^0_{i,j} > \text{Threshold},$$

$$C_{ij} = \sum_k M_{i,k}, \quad \text{if } i = j$$

The suitably changed effective form of GNM Hamiltonian can therefore be written as:

$$H_{\text{GNM}} = \sum_i \frac{(\vec{p}_i)^2}{2m} + \frac{k}{2} \sum_{i,j} C_{i,j} \Delta \vec{R}_i \Delta \vec{R}_j$$

which is, not much of an improvement, but at least it describes the system in exactly the way it is supposed to be viewed. Since residue fluctuations are isotropic, (viz., $\langle (\Delta X_i)^2 \rangle = \langle (\Delta Y_i)^2 \rangle = \langle (\Delta Z_i)^2 \rangle = \langle (\Delta R_i)^2 \rangle / 3$, the position vector $\vec{R}_i$ can be broken up symmetrically in three components (X_i, Y_i, Z_i). Whereupon, to cut down on degeneracy, one can study the behavior of the fluctuation part of the Hamiltonian by studying anyone of these components. Causing no loss to

generality, we choose to study the behavior of only the X-component. Thus we
have the (reduced) Hamiltonian: $H_{\text{GNM}} = \sum_i \frac{(\vec{p_i})^2}{2m} + \frac{k}{2} \sum_{i,j} C_{i,j} \Delta \vec{X}_i \Delta \vec{X}_j$

In an equivalent form, this system can be expressed with Lagrangian description also (Reuveni 2008), whereby we arrive at:

$$L_{\text{GNM}} = \sum_i \frac{M \Delta \dot{X}_i^2}{2} - \frac{k}{2} \sum_{i,j} C_{i,j} \Delta \vec{X}_i \Delta \vec{X}_j$$

Then, using the Euler–Lagrange equation of motion, viz. $\left(\frac{\partial L_{\text{GNM}}}{\partial \Delta X_i} = \frac{\partial}{\partial t} \frac{\partial L_{\text{GNM}}}{\partial \Delta \dot{X}_i} \right)$
and re-arranging the expression, we obtain: $\Delta \ddot{X}_i = -k \sum_j C_{i,j} \Delta X_j$. It assumes
significance to observe here that the last equation implies that dynamics of ith
node, captured in the term ΔX_i, is not only a function of itself but also of the
dynamics of other nodes, described in $\{\Delta X_j\}$. Anyway, upon performing the
summation, this expression can be written in the composite matrix form, as:
$M \Delta \vec{X} = -kC \Delta \vec{X}$. The last equation is a well-known and adhering to the standard technique one may substitute $\Delta \vec{X} = \vec{A} e^{i\omega t}$, to obtain a harmonic solution
of the equation. The eigenvalue equation for the matrix can therefore be given
by: $C \vec{A} = \frac{1}{k} M \omega^2 \vec{A}$;—which implies that for such a description of the system,
eigenfrequencies of the system are, to a proportionality factor, square root of
the eigenvalues of the matrix C. Furthermore, since the spectral dimension
attempts to characterize the *scaling* between density of modes ($g(\omega)$) and mode
frequency (ω), proportionality factor does not play a decisive role here. The set
of vibrational eigenfrequencies (ω_0, ω_1, ..., ω_{N-1}) (alternatively called the 'harmonic spectrum') can subsequently be obtained by diagonalizing the matrix 'C'.
Finally, by plotting $\ln(G(\omega))$ versus $\ln(\omega)$, one can identify the linear stretch in
the ordinate.

How to identify the linear stretch?—Many prefer to achieve this by manu-
ally inspecting every case. But this is tiresome, faulty and for a big dataset,
unrealistic. Instead, consider a window of five consecutive points, and run this
window in an overlapping mode across the entire ordinate profile. For each
of these windows calculate the cumulative difference, viz. $\sum_{i=1}^{5} (y_{i+1} - y_i)$.
The window with least cumulative difference is the best linear stretch for
$\ln(G(\omega))$ versus $\ln(\omega)$ plot for the particular protein under consideration.
Consider the third point of this window and this magnitude of ordinate should
be assigned as the SD of that protein. (Of course, this goes without saying that
same technique to identify the linear stretches should also be applied to all
forms of MFD calculations too.)

Magnitude of SD is generally smaller than that of the MFD of the object
under consideration. This fact relates to the connectivity or bonding of the atoms
(Ben-Avraham 1993) in the object. When SD magnitudes for some small proteins were obtained from results of the spin echo experiments, these were found
to range between 1.3 and 1.6 (Stapleton et al. 1980; Drews et al. 1990). These
magnitudes matched the ones obtained from theoretical and computational analysis on fractal models of proteins (Elber 1989; Elber and Karplus 1986). In a 2004

study on 58 proteins (ranging from less than 100 to 3,600 residues), Burioni and co-workers used the GNM and computed the SD directly from the density of states for these proteins (Burioni et al. 2004). The SD was found to range from about 1.3 to 2.0; furthermore, the SD was reported to increase logarithmically with protein size (Burioni et al. 2004). This last fact assumes significance, because it provides a means to estimate the spectral dimension of a protein based on its size. On the other hand, the fact that SD magnitudes are predominantly found to be less than 2.0, suggested the presence of large thermal fluctuations of residue displacements about their equilibrium position. We will come back to this point later.

2.2.5.2 Interpretations of Spectral Dimension

When Alexander and Orbach (Alexander and Orbach 1982) introduced the spectral dimension, it was considered as a useful tool to characterize the low-frequency vibration spectrum of fractals. Present scenario is different. In this day and age, the spectral dimension is considered by many as the appropriate generalization of the Euclidean dimension of regular lattices to irregular structures in general, whether fractal or not. These irregular structures include polymers, proteins, glasses, percolation clusters, dendritic growths, etc., although we will only talk about spectral dimension in the context of proteins.

Probably, the most tangible interpretation of spectral dimension can be found from the extent of connectivity in a protein; a high magnitude of spectral dimension corresponds to high degree of topological connectedness. Spectral dimension finds relevance in (anomalous) diffusion studies, energy conductivity studies, phase-transition studies, etc. Though various aspects of pertinence of spectral dimension in the paradigm of studies on density of vibrational modes in proteins have already been presented in the last sections, no discussion can be complete without mentioning spectral dimension's pivotal role in derivation of generalized Landau-Peierls instability. We will tangentially touch upon this topic. Although spectral dimension can relate to large-scale topology of protein geometry, it governs the low-energy fluctuations of proteins too. Thus, exploiting this fact, one can derive an instability criterion for proteins, which is based only on topological considerations, which is analogous to Peierls' criterion deduced originally for ordered crystalline structures. We will come back to this topic later.

Finally, it will be incorrect to think that spectral dimension can only relate to various aspects of energy flow and vibrational dynamics, spectral dimension can be useful in predicting dynamical relaxation properties too. Defining vibrational excitation of an object with fractal geometry as 'fractons', a 2005 study (Granek and Klafter 2005) has proposed a model of protein vibrations (Nakayama et al. 1994) to describe fluctuations between pairs of residues. Remarkably, the theoretically predicted results matched well with results obtained from single-molecule experiments (Yang et al. 2003).

2.3 Results Obtained with Fractal Dimension-Based Investigations

It can be realized without much difficulty that fractal dimension-based procedures can extract valuable information about the latent symmetry in arrangement of atoms and residues. Moreover, they can be indispensable in extracting patterns from a spectrum of nonlinear, time-dependent and context-dependent interactions amongst the interior biophysical properties. Thus, not surprisingly, these algorithms have been applied (meaningfully) in a wide range of cases. Due to the aperiodic arrangement of the amino acids, conformational states of proteins consist of many sub-states with nearly the same energy (Frauenfelder et al. 1988). Experimental evidences do indicate that each sub-state of the protein has in itself a large number of sub-states, and the potential energy function is statistically self-similar, having the same form on many different scales (Li et al. 1990a, b). To investigate a system with such characteristics, FD-based constructs present themselves as able candidates. Here we talk about the principal findings from variegated gamut of FD-based results.

2.3.1 FD-Based Protein Conformation Studies

FD is a reliable indicator of protein conformation, because it provides a quantitative measure of the degree to which a structure (and any property associated with the structure) fills the space. Based on this recognition, way back in 1982, Stapleton and his co-workers (Allen et al. 1982) had shown that changes in protein structure with solvent conditions could conveniently be monitored by it. In fact, in its totality, their studies (Stapleton et al. 1980; Allen et al. 1982; Wagner et al. 1985; Colvin and Stapleton 1985) had reported for the first time that the geometry of the carbon backbone determined from X-ray diffraction data and the vibrational dynamics of proteins as measured by Raman scattering, are both fractals. Their work proved that FD determined from the structure, correlates (almost) perfectly with that determined from the dynamics. Subsequently in 1984, another group took this question up and worked further on the Stapleton finding that the FD of the polymer coincides with SD that governs the phonon density of the states $g(\omega)$ at small frequencies (ω) as : $g(\omega) \sim (\omega)^{SD-1}$. However, their interpretation (Helman et al. 1984) of 'fracton's was challenged (Herrmann 1986) soon after (on the ground that fracton dimension only considers the scalar excitations, while the phonons are vector excitations) and exponent in Stapleton finding was related to spectral dimension. For an in-depth account of these concepts, interested readers can refer to (Colvin and Stapleton 1985; Helman et al. 1984; Herrmann 1986; Stapleton 1985) and references therein), to appreciate the pivotal significance of these early works in the context of contemporary theoretical soft condensed matter research.

2.3.2 FD-Based Ion-Channel Kinetics Studies

Liebovitch and co-workers proposed a framework for fractal model of ion channel kinetics in 1987 (Liebovitch et al. 1987a, b; Liebovitch and Sullivan 1987). The fact that their model was more consistent with the conformational dynamics of proteins, helped in establishing the fact that fractal models can be accurate and reliable when considering conformational transitions between states that consist of a hierarchy of sub-states. Liebovitch's interpretation of the results was radical, because it was based on the notion that dynamic processes in proteins occur with many different correlation times. By adjusting the parameters of the model, the observed gating behavior could be described in a wide time range. The immediate impact of this study on the then-contemporary ones could be understood by comparing the (varying) approaches and interpretations suggested in (Korn and Horn 1988; French and Stockbridge 1988; Millhauser et al. 1988; Lauger 1988), with those that were attempting to describe the same genre of results with other tools. Subsequent works by Liebovitch (Liebovitch and Toth 1990; Liebovitch and Toth 1991) on analysis of patch clamp recordings of the sequence of open and closed times of cell membrane ion channels, could establish a new trend ion channel kinetics studies, which asserted that ion channel proteins have many conformational states (connected by large number of pathways) of nearly equal energy minima. It was shown that these states are not independent but are linked by physical mechanisms that result in the observed fractal scaling. Impact of these findings can be appreciated from the profile of succeeding studies (Lowen and Teich 1993; Churilla et al. 1995; Lowen et al. 1999; Rodriguez et al. 2003; Kim et al. 2005) on fractal kinetics from 1991 onwards. For example, in a recent work (Kim et al. 2005) it has been shown that patterns in spiking activity in suprachiasmatic nucleus neurons can be explained with fractal point process model, which implied the presence of self-similarity in the profile of spike trains; proving thereby that no characteristic time scales dominates the dynamics of the spiking process.

2.3.3 FD-Based Attempts to Relate Protein Structure and Dynamics

It was known from a long time that protein dynamics involves a broad range of interconnected events occurring on various time scales (Brooks et al. 1988). Thus, it is hardly surprising that protein systems exhibit nonexponential behavior. Fractal approach is of great assistance in developing models of protein dynamical processes. There is a wealth of nonexponential rate processes in biophysics, and fractal models were found to be effective in describing such processes (Dewey and Spencer 1991). Although some dynamical models (Bagchi and Fleming 1990; Zwanzig 1990) were there to describe the vast array of nonexponential processes, discriminating between them was a difficult (yet important) problem. In this context, fractal dimension-based models came handy. Fractal models had a unique appeal because with them a structural parameter, the MFD, could (automatically) be related to a dynamic parameter, viz. SD, the spectral dimension. Exploiting this advantageous aspect, Dewey (Dewey and Spencer 1991; Dewey and Bann 1992; Dewey 1995) explored the inherent direct connection between structure and dynamics in a series of works. To study the emergence and characteristics of these nonexponential behaviors in conformational change of proteins and protein structures in general, one can refer to (Dewey 1997). Recent applications of this approach can be found in a number of works (Ramakrishnan and Sadana 1999; Goychuk and Hanggi 2002; Carlini et al. 2002).

2.3.4 Studies in Fractal Kinetics

Discussion above (on fractal studies of ion channel kinetics and nonexponential rate processes) falls under the broad category of "fractal kinetics" study. It was Kopelman (1986), who suggested that the general phenomenon of logarithmically decreasing reaction rates should be collectively considered under the umbrella of "fractal kinetics". His assertions were proved experimentally in a number of boundary conditions (Kopelman 1988). Spatial and/or energetic heterogeneity of the medium, or nonrandomness of the reactant distribution in low dimensions was suggested as (possible) mechanisms behind fractal kinetics. Kopelman (1988) showed how surface diffusion-controlled reactions, which occur on clusters or islands, are expected to exhibit anomalous and fractal-like kinetics. Mechanisms of fractal kinetics exhibit anomalous reaction orders and time-dependent (for example, binding) rate coefficients, holding therefore an advantageous ground. Investigations (Li et al. 1990a, b) (with theoretical considerations raised by statistical methods of random walk on fractal structures) have proved that fractal kinetics is observed not due to the number of binding sites on each protein molecule, but it is a reflection of the fractal properties of enzymes (and proteins in general) themselves. Since roughness of protein surfaces follow fractal scaling (Lewis and Rees 1985) and since fractal dimension is a global, statistical property, insensitive to time-dependent tiny fluctuations;—the results in (Li et al. 1990a, b) appear to be easily understandable. The same study had (notably) established that, nonintegral dimensions of the Hill coefficient (used to describe the allosteric effects of proteins and enzymes), are direct consequences of fractal properties of proteins. Another consequence of this inherent fractal nature was demonstrated by power-law dependence of a correlation function on certain coordinates, and power-law dependence of correlation function on the surface of analyte-receptor complex on time (Ramakrishnan and Sadana 1999). Original Kopelman algorithm was provided a strong theoretical support when it was shown that fractal kinetics owes its emergence to spatial self-organization of the reactants induced by the compact properties of diffusion (Argyrakis and Kopelman 1990). Being an extremely significant platform, countless applications of this scheme have been attempted over the years; one can find the prominent ones in (Turner et al. 2004; Berry 2002; Yuste et al. 2004; Kosmidis et al. 2003; Grimaa and Schnell 2006). Interested reader may also benefit from studying (Turner et al. 2004), where an in-depth comparison between various pertinent schemes is provided.

A discussion (however small) on fractal kinetics cannot be complete without a note on fractal time studies. However, such a note will fall outside the purview of the present book. Interested reader can find an excellent discussion about chemical kinetics and fractal time and their pertinence in the realm of protein interaction studies elsewhere (Shlesinger 1988), pages 158–163 of (Dewey 1997).

2.3.5 FD-Based Results on Protein Structure

The recent trend of FD-based studies indicates a strong inclination toward discovering protein structural invariants. Since MFD is a measure of the compactness, it can alternatively be used as an effective measure for the extent of torsion that the extended polypeptide undergoes during folding. It is expected (intuitively) that the stronger the torsion of the systems is, the larger will be its MFD, and vice versa. Therefore to quite an extent, MFD reflects the structure information of proteins, and submits itself as a reliable construct to study the extent of folding. Enright and Leitner in 2005 (Enright and Leitner 2005) formalized the aforementioned intuitive notion by systematically calculating the MFD of a set of 200 proteins and found the magnitude to be 2.489 ± 0.172 . Their result was consistent with those obtained from dispersion relations and the anomalous subdiffusion, computed for the same set of proteins (Yu and Leitner 2003), which were (in turn) related to the theory of mass fractal dimension proposed by Alexander and Orbach (Alexander and Orbach 1982). A lot of incisive research works in the last 5 years have zoomed in on two aspects of Enright–Leitner finding; one, influence of magnitude of the MFD on protein dynamics and energy flow, and two the noncompactness of the folded primary structure. Many findings obtained from these studies can be labeled "unexpected", because the classical works on protein structure could not reveal them. Here we outline some such results.

The similarity between the geometry of a globular protein and a percolation cluster was pointed out by Liang and Dill (Liang and Dill 2001). However, based on a (masterly) follow-up study (Leitner 2008), it was found out that energy flow through the vibrational states of proteins is anisotropic, and this process can be reliably described by the theory of transport on a percolation cluster in terms of the dimension of the cluster and the density of vibrational states. Though the precise functional roles of the energy propagation channels could not be identified; possibility of relating them to fast cooling of proteins during reaction, directed energy transport to speed reaction, communication signals, and allostery—were proposed.

Adopting the fractal point of view made it possible to analyze protein topology and protein dynamics, within the same framework. This advantageous fact was utilized wonderfully in a recent work (de Leeuw et al. 2009) where a universal equation of state for protein topology was derived. Although the marginal stability criterion of proteins was talked about previously (Li et al. 2004), this study (de Leeuw et al. 2009) proved that most proteins "exploit" the Landau-Peierls instability (Peierls 1934) to attain large amplitude vibrations. The authors insisted that this is the key mechanism how proteins can perform large amplitude conformational changes to ensure proper function, while maintaining an invariant native fold. Furthermore (perhaps most importantly), it is proved here that the majority of proteins in the PDB exist in a marginally stable thermodynamic state, namely a state that is close to the edge of unfolding. These are exciting new findings about state of existence of proteins, which could never have been brought to the fore without efficient and accurate implementation of FD-based ideas.

Recently, the horizon in this area opened up further when the answer to a basic question, namely, 'why, although proteins seem to vibrate under a complex energetic landscape thanks to their highly involved dynamics, the same energetic landscape appear harmonic near the minima?'—was answered. This was achieved (Reuveni et al. 2010) by challenging the very validity of the harmonic approximation concerned and the study could establish that anomalous vibrational dynamics in proteins is a manifestation of a fractal-like native state structure. To obtain an even more inclusive picture of this aspect of protein structure studies, one can complement the aforementioned (Reuveni et al. 2010) assertion with that from another recent study (Lois et al. 2010), where the later could identify a possible mechanism, the small fractal dimension of "first passage networks", that accounts for reliable folding for proteins with rugged energy landscapes. On a related note, one may find it useful to note the results presented in a recent work on folding of proteins that are too slow to be simulated directly. Here, it was asserted that fractal Smoluchowski equation could reliably model protein folding by subdiffusion (Sangha and Keyes 2009). In this way, this methodology could achieve what the standard paradigm of, either diffusion (the usual Smoluchowski equation), or normal-diffusion continuous time random walk of a single-order parameter under thermodynamic influences,—would have had difficulty to achieve.

Another equally unexpected set of results (Banerji and Ghosh 2009b) was presented when the degrees of noncompactness of mass, hydrophobicity, and polarizability within proteins were quantified across four major structural classes for globular protein structures. This study revealed that barring all-α class, all the major structural classes of proteins have an amount of unused hydrophobicity left in them. Figure 2.2 describes the presence of unused hydrophobicity in bacterial family-III cellulose-binding domain protein `1nbc' (all-beta domains). Amount of unused hydrophobicity was observed to be greater in thermophilic proteins, than that in their (structurally aligned) mesophilic counterparts. The all-β proteins (thermophilic, mesophilic alike) were identified with maximum amount of unused hydrophobicity, while all-α proteins have been found to possess minimum polarizability. The unused hydrophobicity was explained as the extent of hydrophobicity that was not utilized during the process of protein folding. It was hypothesized in the work that such unused hydrophobicity acts as an added layer of protection for proteins, lending them with an extra level of stability. The fact that proteins

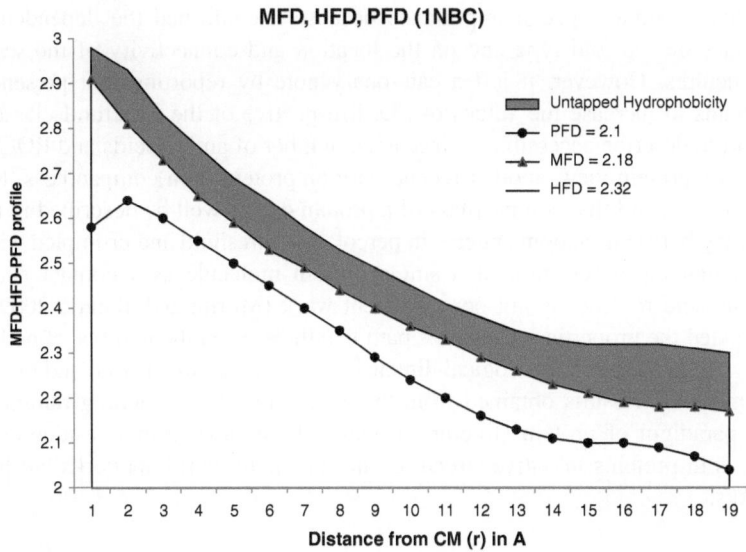

Fig. 2.2 The mass fractal dimension, hydrophobicity fractal dimension and polarizability fractal dimension of an 'all-beta' class protein (PDB id.: 1NBC)

belonging to thermophilic organisms have more unused hydrophobicity in them, therefore, does not appear strange. Polarizability, on the other hand, relates to the bottom-up description of protein structure, because it concerns the residue-specific magnitudes. Lower profile of residue polarizability implies lower dielectric constant in protein interior; which in turn, suggests a conducive electrostatic environment in protein interior. Interpreting the obtained findings from the afore-mentioned perspectives, the Banerji–Ghosh work (Banerji and Ghosh 2009b) thus asserted that origin of α-helices are possibly not hydrophobic but electrostatic, whereas β-sheets are hydrophobic in nature.

Staying with noncompactness, approaching the questions from a different per-spective, another recent work studied the backbone of ribosome (Lee 2008), employ-ing correlation dimension (discussed earlier). It revealed that capacity dimensions of rRNAs are less than their embedding dimension, implying presence of some empty spaces in rRNA backbones. Going back to principles of polymer physics in the context of proteins, Hong and Jinzhi (2009) derived a unified formula for the scal-ing exponent of proteins under different solvent conditions. Since this formula was obtained by considering the balance between the excluded volume effect and elastic interactions among monomers, one can expect it to find a lot of attention in pure and applied research works of future. Their results showed that the scaling exponent is closely related to the MFD of a protein's structure at the equilibrium state.

Hidden dependencies between protein structural class-specific FD mag-nitudes and kinetic-thermodynamic parameters (the folding and unfolding rate, folding-unfolding free energy) were studied recently (Tejera et al. 2009).

This study assumes special importance because it confirmed the dependence of FD values on the fold type and on the location and connectivity of the secondary structures. However, it left a cautionary note by reporting that presence of turns tends to increase the values of FD, irrespective of the structural classes. In an effort to describe accessible surface area, number of amino acids and ROG with FD-centric protein view, another recent work on protein (non)compactness (Moret et al. 2009), found that compactness of a protein can as well be described in terms of packing between random spheres in percolation threshold and crumpled wires.

Although the description of a single protein molecule as a contact network of amino-acid residues is not new; a recent work (Morita and Takano 2009) has investigated the properties of shortest path length between the number of nodes by employing the network topological dimension, the fractal dimension, and the spectral dimension. Results obtained from this study are of far-reaching ramification in the paradigm of protein structure, because these showed that residue contact networks in proteins in native structures are not small world networks but fractal networks.

2.4 Gaining New Knowledge About Protein Interior with FD-Based Investigations

While the last section enlists various successful applications of FD-based constructs in probing protein interiors, a lot can still be achieved in that end. Just to emphasize this very point, that is, how, yet-unexplored questions can be rigorously investigated with innovative use of FD constructs; I present here a glimpse of correlation-dimension (CD)-based investigations of protein structural parameters. Since one can investigate the underlying symmetries in short, medium, and long-range interactions between various residues with an invariant framework, CD-based studies on protein interior, can prove to be extremely useful. The scope and resolution of the obtained results underscore the point that, extensive information about numerous (elusive) dependencies between biophysical and/or biochemical invariants within proteins, can be drawn from series of elaborate and systematic studies—using this very methodology.

2.4.1 Implementation of CD-Based Constructs

CD (as explained earlier) is different than the other FD constructs, because it attempts to investigate the self-similarity in distributions between correlated properties (or dependencies), which arise out of particular dispositions of atoms or residues. In other words, instead of studying the self-similarity in distribution of atoms or residues—something that the mass-fractal analyses perform; here, one examines the self-similar characteristics of correlations amongst the properties that come to being due to any given distribution of atoms or residues. Figures 2.3, 2.4, 2.5 , and 2.6 contain correlation-dimension-based analysis of key biophysical properties necessary for protein stabilization. While some of these are quantifiable through other possible means; some say the possible self-similarity in the distribution of π-electron clouds of the aromatic amino acids, will be difficult to obtain

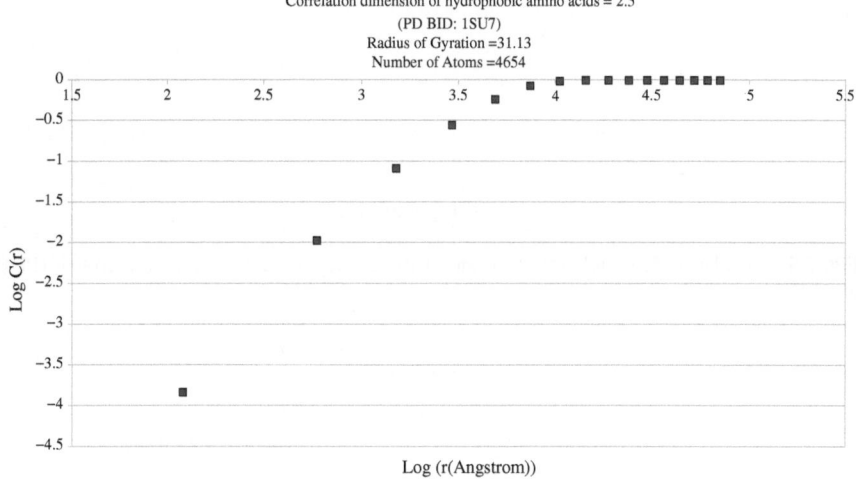

Fig. 2.3 Calculation of correlation dimension between hydrophobic amino acids in a protein (PDB id.:1SU7)

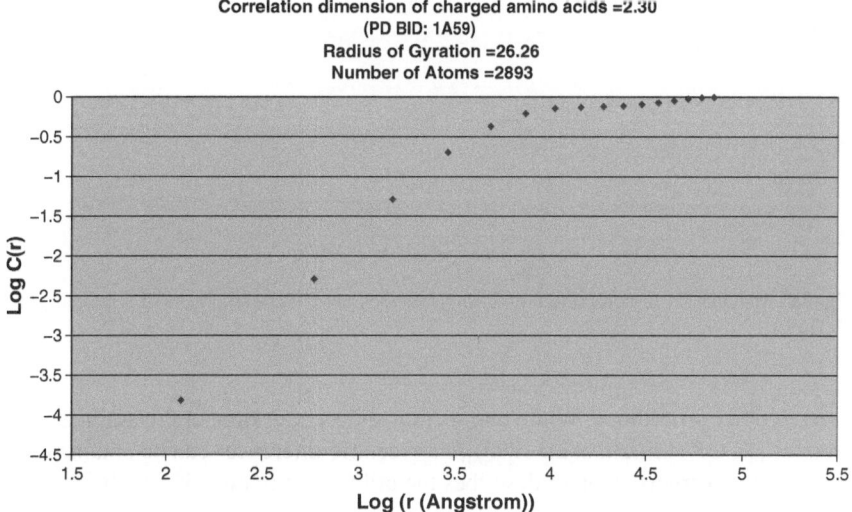

Fig. 2.4 Calculation of correlation dimension between charged amino acids in a protein (PDB id.:1A59)

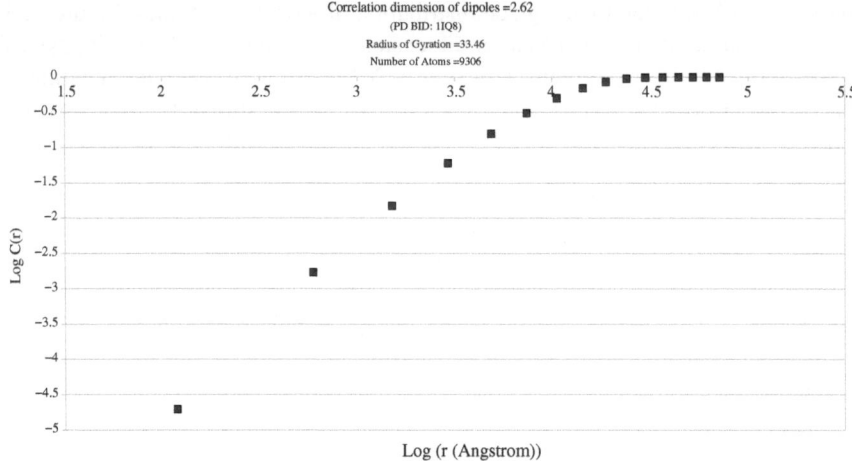

Fig. 2.5 Calculation of correlation dimension between peptide dipole units in a protein (PDB id.:1LQ8)

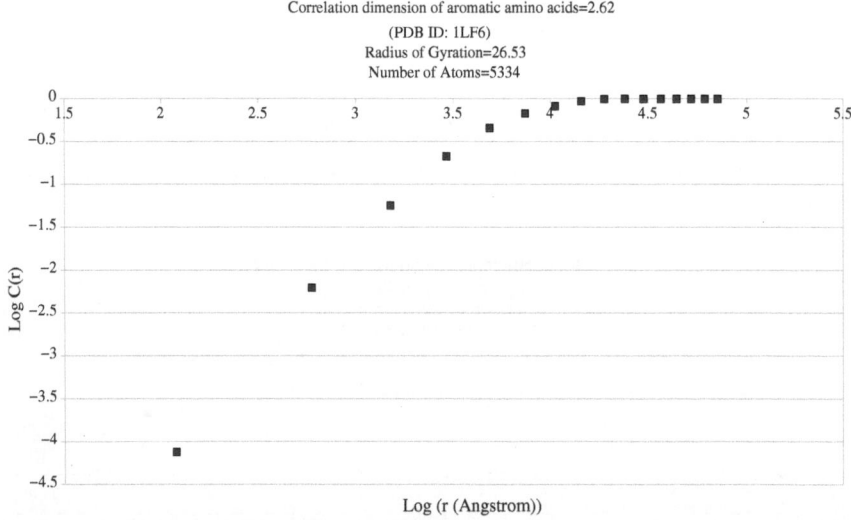

Fig. 2.6 Calculation of correlation dimension between aromatic amino acids in a protein (PDB id.:1LF6)

through other procedures. While one section shows how structural classification of proteins [SCOP (Murzin et al. 1995)]-class centric inferences can be obtained with CD-based constructs, another describes the utility of it in quantifying the extent of dependencies amongst properties of individual residues. Thus, while one section, presents results of general nature; another presents the results at the resolution of particular proteins.

2.4.2   CD-Based Investigation of Local and Global Dependencies Amongst Peptide Dipoles Units

2.4.2.1   Why is it Advantageous to Resort to CD, While Studying Protein Electrostatics?

Many quantitative analyses have attempted to describe the delicate balance of forces in the complex charge subsystem and hard-to-model dielectrics of protein interior. Proposed simulation strategies to study this problem range from continuum dielectric methods (Bashford and Karplus 1990) to explicit approach for dealing with polarizability (Warshel and Papazyan 1998). In other words, starting from Tanford–Kirkwood dielectric cavity scheme, analytical solution of Poisson–Boltzmann equation, nonlinear numerical finite difference (Zhou et al. 1996), boundary element algorithms (Lu et al. 2005), and finally the sophisticated empirical generalized Born solutions (Feig et al. 2004)—all have been tried. However, explicit treatment of polarizability, even at its modest linear response level, involves high-computational price and detailed know-how of molecular modeling practices. Thus, the immediate needs of many protein scientists, viz. easy-to-understand and easy-to-calculate measures to describe protein electrostatics, are not always addressed by the aforementioned procedures. In this context, one might find the use of CD-based analyses of protein interior to be immensely beneficial. Since charge–charge and peptide–dipole interactions have been identified to play important roles in the stabilization of proteins (Spassov et al. 1997) and in determining the native fold (Petrey and Honig 2000), here I demonstrate how CD-based investigations on these can bring to fore valuable information from an invariant simple framework.

2.4.2.2   Calculation of CD to Quantify Dipole Distribution

Although the peptide group is uncharged at normal pH, its double-bonded resonance form accounts for a dipole moment of ~3.6 Debye, directed from oxygen of C=O, to hydrogen of N–H bond. In case of some secondary structures, like the α-helices, these dipole moments align with each other to produce a significant resultant dipole moment, which in turn, tend to account for nontrivial contributions to protein stability. Electrostatic field resulting from arrangement of helix dipoles has been proposed as an important factor contributing to both helix stability and specificity (Shoemaker et al. 1987). Aqvist et al. (1991) have demonstrated that this effect is short ranged, and its influence is confined largely to individual backbone dipoles localized within the first and last turns of the helix resulting in a formal positive charge at the helix N terminus and formal negative charge at the C terminus. But the global picture regarding protein interior electrostatics, as a function of backbone conformation of the native structure—does not, necessarily, emerge from that. Since dipoles in an α-helix are electrostatically aligned nearly parallel to the axis of the helix (Wada 1976), such a global picture can be obtained straightaway by calculating the general dependency profile between dipoles, originating from the numerous mutual spatial dependencies between the peptide-unit dipoles across the entire backbone. The mere vectorial summation to calculate the resultant dipole vector will fall short of quantifying this aspect of dipole distribution in protein structures.

Hence, choosing the standard threshold for considering a pair of residues to be in long-range contact (Miyazawa and Jernigan 1999; Selvaraj and Gromiha 2003), namely a maximal distance of 8 Å between their centroids (alternatively, distance between C^β atoms (C^α for glycines) can be chosen, results do not vary appreciably (unpublished data)) and resorting to the formulae provided earlier, CD were calculated for every nonredundant protein structure (taken from protein data bank [PDB (Berman et al. 2003)] within all the SCOP folds with statistically significant number of nonredundant protein structures. While calculating the centroid of an amino acid, all the nonhydrogen atoms were considered. Set of formulae mentioned earlier was used to calculate the correlation dimension. One can refer to (Banerji and Ghosh 2011) for details of the methodology. After obtaining the CD for each of the proteins belonging to a particular SCOP fold, average of obtained CDs for a particular SCOP fold was subsequently calculated to draw biological inferences at the SCOP class level. Tables 2.1 and 2.2, contain the list of 15 SCOP folds with respectively the maximum and minimum magnitudes of CD between peptide dipoles, the entire list can be found in (Banerji and Ghosh 2011).

Brief Discussion About Peptide–Dipole Distribution Within SCOP-Folds

Most of the studies on peptide–dipole units (Shoemaker et al. 1987; Wada 1976) have concentrated on the nature and magnitude of (resultant) dipoles moments, especially in the context of α-helices. Results obtained with CD, on the other hand, quantified a different aspect of peptide–dipole units in proteins; namely, it quantified the prevalent symmetry in dependencies amongst the spatial arrangement of the peptide dipoles, across proteins from all the seven SCOP classes. Somewhat unexpectedly, the results revealed that peptide dipoles in α/β class of proteins, in general, are more correlated to each other, than the peptide dipoles in proteins belonging to all-α class. Out of 15 SCOP folds with maximum CD amongst peptide dipoles (Table 2.1), only one ('α/α toroid', with the maximum CD) entry belongs to all-α class, eight belongs to α/β class of proteins, three belongs to $\alpha + \beta$ proteins, along with two entries from all-β and one from multidomain proteins. Hence, although peptide dipoles in α-helices are electrostatically aligned nearly parallel to the axis of the helix (Wada 1976) and although their resultant effect in helix stability and specificity (Shoemaker et al. 1987), is well documented; with respect to ensuring maximum dipole–dipole interactions, the α/β structures appear to be better optimized. On the other hand, in the particular case of fold 'α/α toroid', the closed circular arrangement of array of α-hairpins, present perfect positioning for the peptide dipoles to correlate with each other; which explain α/α toroid's having highest CD for peptide dipoles.

Table 2.1 List of 15 SCOP-folds with maximum correlation-dimensions between peptide unit dipoles

SCOP class	SCOP fold	Magnitude of correlation dimension
All-α proteins	α/α toroid	2.490
α/β proteins	PLP-dependent transferase-like	2.469
Multi-domain protein	β-lactamase/transpeptidase-like	2.463
α/β proteins	α/β -Hydrolases	2.452
$\alpha + \beta$ proteins	Protein kinase-like (PK-like)	2.432
α/β proteins	TIM β/α -barrel	2.428
α/β proteins	Ribokinase-like	2.426
α/β proteins	Nucleotide-diphospho-sugar transferases	2.407
α/β proteins	Periplasmic binding protein-like II	2.379
α/β proteins	S-adenosyl-L-methionine-dependent methyl transferases	2.368
$\alpha + \beta$ proteins	Class II aaRS and biotin synthetases	2.363
$\alpha + \beta$ proteins	Cysteine proteinases	2.342
α/β proteins	Phosphorylase/hydrolase-like	2.337
α/β proteins	HAD-like	2.327
$\alpha + \beta$ proteins	Ntn hydrolase-like	2.325

Table 2.2 List of 15 SCOP-folds with minimum correlation-dimensions between peptide unit dipoles

SCOP class	SCOP fold	Magnitude of correlation dimension
Peptides	Conotoxins	0.539
Peptides	Transmembrane helical fragments	0.933
Coiled coil proteins	Parallel coiled coil	1.095
Designed proteins	Zinc finger design	1.125
Small proteins	Knottins (small inhibitors, toxins, lectins)	1.252
Small proteins	β-β-α zinc fingers	1.309
Small proteins	Glucocorticoid receptor-like (DNA-binding domain)	1.360
Coiled coil proteins	Stalk segment of viral fusion proteins	1.431
Small proteins	Rubredoxin-like	1.512
All-α proteins	RuvA C terminal domain like	1.628
All-α proteins	Long-α hairpin	1.675
All-β proteins	Glycosyl hydrolase domain	1.730
All-β proteins	SH3 like barrel	1.768
All-α proteins	λ-repressor like DNA binding domains	1.834
All-α proteins	DNA RNA binding 3 helical bundle	1.854

The assertion about better inherent correlations between dipoles in α/β proteins is vindicated by Table 2.3 and by results enlisted in Table 2.2 too. Amongst the 15 SCOP folds with minimum CD between peptide dipoles, not a single entry

Table 2.3 Correlation dimension between peptide unit dipoles, across four major SCOP classes

SCOP class	Dipole moment correlation dimension (mean ± standard deviation)
α/β proteins	(0.958 ± 0.252) – highest
$\alpha + \beta$ proteins	(0.725 ± 0.220)
All-α proteins	(0.629 ± 0.241)
All-β proteins	(0.624 ± 0.258) – least

could be found from α/β proteins. Four entries from 'small proteins' class and one each from 'peptides' and 'designed proteins' class, underline the fact that in these SCOP folds, the correlation (and dependencies) between peptide–dipole units on each other is extremely less. On the other hand, Table 2.3 results show unambiguously that spatial correlation (and therefore, dependency) between peptide dipoles in all-α and all-β class of proteins match each other almost perfectly.

2.4.3 CD Amongst Charged Amino Acids Across All Seven SCOP Classes

Considering 'His', 'Arg', 'Lys', 'Asp' and 'Glu' as the charged amino acids, analysis was carried out on the nonredundant structures from all the seven SCOP classes with unchanged methodology as used in the previous case. Table 2.4 enlists 15 SCOP folds with maximum values of CD amongst charged amino acids, Table 2.5 enlists

Table 2.4 List of SCOP folds with maximum correlation-dimensions amongst the charged amino acids ('His', 'Arg', 'Lys', 'Asp' and Glu')

SCOP class	SCOP fold	Magnitude of correlation dimension
$\alpha + \beta$ proteins	Class II aaRS and biotin synthetases	2.324
$\alpha + \beta$ proteins	Protein kinase-like	2.322
$\alpha + \beta$ proteins	Lysozyme-like	2.275
α/β proteins	PLP-dependent transferase-like	2.273
All-α proteins	α/α toroid	2.261
α/β proteins	HAD-like	2.258
Multidomain proteins	β-lactamase/transpeptidase-like	2.253
$\alpha + \beta$ proteins	Zincin-like	2.251
$\alpha + \beta$ proteins	Cysteine proteinases	2.241
$\alpha + \beta$ proteins	ATP-grasp	2.240
α/β proteins	Nucleotide-diphospho-sugar transferases	2.239
α/β proteins	TIM β/α–barrel	2.238
α/β proteins	S-adenosyl-L-methionine-dependent methyltransferases	2.234
α/β proteins	α/β -Hydrolases	2.212
α/β proteins	Ribokinase-like	2.208

Table 2.5 List of SCOP folds with minimum correlation-dimensions amongst the charged amino acids ('His', 'Arg', 'Lys', 'Asp' and Glu')

SCOP class	SCOP fold	Magnitude of correlation dimension
Peptides	Conotoxins	0.920
Peptides	Transmembrane helical fragments	1.250
Coiled coil proteins	Stalk segment of viral fusion proteins	1.488
Small proteins	Glucocorticoid receptor-like (DNA-binding domain)	1.606
Coiled coil proteins	Parallel coiled coil	1.628
Small proteins	Knottins (small inhibitors, toxins, lectins)	1.711
Designed proteins	Zinc finger design	1.729
All-β proteins	Glycosyl hydrolase domain	1.769
All-α proteins	RuvA C terminal domain like	1.863
Small proteins	β-β-α zinc fingers	1.887
Small proteins	Rubredoxin-like	1.929
All-α proteins	Long-α hairpin	1.949
All-β proteins	Immunoglobulin like β-sandwich	1.969
$\alpha + \beta$ proteins	FKBP-like	1.973
All-α proteins	λ-repressor like DNA binding domains	1.980

15 SCOP folds with minimum magnitudes of CD amongst charged amino acids. Interestingly, six out of top ten correlation dimensions between charged amino acids could be observed in proteins from α/β class (Table 2.4). This is in striking similarity with the results obtained from SCOP class wide peptide–dipole correlation pattern (Table 2.1), where eight of the top fifteen CD magnitudes were observed in proteins from α/β class. The extent of optimization in electrostatic environment of α/β proteins could alternatively be assessed from their (conspicuous) absence from Table 2.5 too. Although it was expected that the proteins from SCOP class 'small proteins' and 'designed proteins' might have minimal CD among the (aforementioned) five charged residues (Table 2.5), it was unexpected to observe that CD among those five charged residues even among the proteins from three all-α folds ('RuvA C terminal domain like', 'λ-repressor like DNA binding domains', and 'SAM domain like') is minimal.

2.4.4   CD Amongst Aromatic Amino Acids Across All Seven SCOP Classes

Considering all the (nonhydrogen) atoms belonging to aromatic rings of 'Tyr', 'Trp', and 'Phe', analysis was carried out on the nonredundant structures from all the seven SCOP classes with unchanged methodology as used in the previous case. Table 2.6 enlists 15 SCOP folds with maximum values of CD amongst π-electron clouds of the aromatic amino acids, whereas Table 2.7 enlists 15 SCOP folds with minimum magnitudes of CD amongst the same. Conforming to the

Table 2.6 List of SCOP, folds with maximum correlation-dimensions amongst the aromatic amino acids

SCOP class	SCOP fold	Magnitude of correlation dimension
Multidomain proteins	β-lactamase/transpeptidase-like	2.531
All-α proteins	α/α toroid	2.497
All-β proteins	β -Trefoil	2.424
All-β proteins	GroES like	2.389
All-β proteins	Trypsin like serine proteases	2.387
$\alpha + \beta$ proteins	Protein kinase-like	2.386
α/β proteins	TIM β/α -barrel	2.381
α/β proteins	α/β -Hydrolases `	2.348
$\alpha + \beta$ proteins	Cysteine proteinases	2.341
α/β proteins	Nucleotide-diphospho-sugar transferases	2.338
$\alpha + \beta$ proteins	ATP-grasp	2.334
α/β proteins	Adenine nucleotide α- hydrolase-like	2.333
α/β proteins	Ribokinase-like	2.328
α/β proteins	PLP-dependent transferase-like	2.325
$\alpha + \beta$ proteins	Ntn hydrolase-like	2.311

Table 2.7 List of SCOP folds with minimum correlation-dimensions amongst the aromatic amino acids

SCOP class	SCOP fold	Magnitude of correlation dimension
Peptides	Conotoxins	0.369
Designed proteins	Zinc finger design	1.512
Small proteins	Rubredoxin-like	1.199
All-α proteins	RuvA C terminal domain like	1.276
Coiled coil proteins	Parallel coiled coil	1.320
All-α proteins	λ-repressor like DNA binding domains	1.408
Small proteins	Glucocorticoid receptor-like (DNA-binding domain)	1.414
All-α proteins	EF hand like	1.445
All-α proteins	Spectrin repeat like	1.498
Peptides	Transmembrane helical fragments	1.506
All-β proteins	SH3 like barrel	1.517
Small proteins	Knottins (small inhibitors, toxins, lectins)	1.544
All-α proteins	SAM domain-like	1.603
Small proteins	β-β-α zinc fingers	1.605
All-α proteins	Four helical up and down bundle	1.616

trends observed in the distribution of peptide dipoles and charged amino acids, six out of top fifteen correlation dimensions between aromatic amino acids could be observed in proteins from α/β class (Table 2.6). Aromatic amino acids from proteins belonging to $\alpha + \beta$ class, were found to feature in the same list for four out of top fifteen cases. Marked absence of proteins belonging to either of these

SCOP classes from Table 2.7 implied the existence of conducive electrostatic environments within α/β and $\alpha + \beta$ proteins in general; so that interactions between π-electron clouds of 'Tyr', 'Trp', and 'Phe' are maximized. On an unexpected note, it was observed that interactions between aromatic amino acids of proteins belonging to SCOP class all-α is nominal amongst all seven SCOP classes, which is demonstrated by the presence of six all-α folds in Table 2.7.

2.4.5 CD Amongst Hydrophobic Amino Acids Across All Seven SCOP Classes

Analysis of self-similarity in interactions amongst hydrophobic amino acids was performed by quantifying their dependencies (or correlations) on each other. This was achieved by calculating the correlation dimensions amongst 'Gly', 'Ala', 'Val', 'Ile', 'Leu', 'Met', and 'Cys', for proteins belonging to seven SCOP classes. The observed pattern that more than half of the SCOP folds with maximum values of CD amongst the hydrophobic amino acids belonged to α/β class (Table 2.8), pointed unambiguously to superior stability profile of proteins belonging to α/β structural class, in general. Such assertion of better stability was further vindicated by a marked absence of α/β folds in Table 2.9, which enlisted 15 SCOP folds with minimum values of CD amongst hydrophobic amino acids. Although 'peptides', 'small proteins', 'designed proteins' were expected to have negligible hydrophobic interaction profiles, presence of four all-α folds ('RuvA C terminal domain like',

Table 2.8 List of SCOP folds with maximum correlation-dimensions amongst the hydrophobic amino acids

SCOP class	SCOP fold	Magnitude of correlation dimension
All-α proteins	α/α toroid	2.358
α/β proteins	α/β -Hydrolases	2.315
α/β proteins	PLP-dependent transferase-like	2.309
Multidomain proteins	β-lactamase/transpeptidase-like	2.295
α/β proteins	TIM β/α -barrel	2.287
α/β proteins	Ribokinase-like	2.285
$\alpha + \beta$ proteins	Protein kinase-like	2.267
α/β proteins	Nucleotide-diphospho-sugar transferases	2.259
$\alpha + \beta$ proteins	Cysteine proteinases	2.242
$\alpha + \beta$ proteins	Class II aaRS and biotin synthetases	2.236
α/β proteins	S-adenosyl-L-methionine-dependent methyltransferases	2.223
All-β proteins	Concanavalin A like lectins glucanases	2.201
α/β proteins	Phosphorylase/hydrolase-like	2.196
All-β proteins	Acid proteases	2.192
α/β proteins	Periplasmic binding protein-like II	2.188

Table 2.9 List of SCOP folds with minimum correlation-dimensions amongst the hydrophobic amino acids

SCOP class	SCOP fold	Magnitude of correlation dimension
Peptides	Conotoxins	0.497
Peptides	Transmembrane helical fragments	1.045
Small proteins	Knottins (small inhibitors, toxins, lectins)	1.103
Designed proteins	Zinc finger design	1.118
Coiled coil proteins	Parallel coiled coil	1.157
Small proteins	β-β-α zinc fingers	1.191
Small proteins	Glucocorticoid receptor-like (DNA-binding domain)	1.305
Small proteins	Rubredoxin-like	1.357
All-α proteins	RuvA C terminal domain like	1.379
Coiled coil proteins	Stalk segment of viral fusion proteins	1.413
All-α proteins	Long-α hairpin	1.498
All-α proteins	λ-repressor like DNA binding domains	1.582
All-β proteins	SH3 like barrel	1.587
All-β proteins	Glycosyl hydrolase domain	1.611
All-α proteins	SAM domain-like	1.614

'long-α hairpin', 'λ-repressor like DNA binding domains', and 'SAM domain like') and two all-β folds ('SH3 like barrel' and 'Glycosyl hydrolase domain') in Table 2.9, indicated that it is simplistic to expect that the nature of hydrophobic interactions in proteins belonging to some all-α and all-β SCOP folds is necessarily different from that existing in 'peptides', 'small proteins', and 'designed proteins'.

2.4.6 General Inferences About Protein Stability from SCOP Class Wide Separate Analyses of CD Between Charged Residues, Peptide Dipole, Aromatic Amino Acids, and Hydrophobic Amino Acids

The CD-based analyses conducted in here involved three aspects of protein electrostatics. Separately, it presented an account of self-similarity in hydrophobic interactions. Although this not exhaustive, even with such limited analyses, certain significant information about latent features of protein stability could be unearthed, which included a detailed characterization of electrostatic environment within protein structures belonging to various SCOP folds across seven SCOP classes. For example, it is interesting to note that 10 out of 15 SCOP folds with maximum CD between 'His', 'Arg', 'Lys', 'Asp' and 'Glu' (Table 2.4), feature within the top 15 SCOP folds with highest magnitudes of CD amongst peptide dipoles (Table 2.1). Even more striking is the fact that the two SCOP folds ('α/α

toroid' (belonging to all-α SCOP class) and 'PLP-dependent transferase-like' (belonging to α/β SCOP class)) with maximum CD between the charged residues and separately, between peptide dipoles—retain their places in respective tables (albeit in reverse order). These signify the presence of extremely favorable electrostatic environments in the proteins belonging to these folds. The fact that some of these SCOP folds with maximal CD magnitudes are amongst evolutionarily most conserved folds (say, TIM β/α-barrel, α/β-hydrolases etc.) therefore, does not appear to be surprising. On a diametrically opposite scenario, all the 10 SCOP folds with minimum CD between 'His', 'Arg', 'Lys', 'Asp' and 'Glu' (Table 2.5), can be observed to feature within the 15 SCOP folds with least magnitudes of CD amongst peptide dipoles (Table 2.2). Proteins from SCOP fold 'Zinc finger design' (under the SCOP class 'Designed proteins') could be identified as ones with least conducive electrostatic environment, since these proteins can be observed to possess the lowest CD amongst charged amino acids and separately, amongst their peptide dipoles. 'Knottins' (small inhibitors, toxins, lectins), under the SCOP class 'small proteins', registered the second position in both the tables for folds with least CD.

2.4.7   Correlation Dimension-Based Investigation of Dependency Distribution Amongst Hydrophobic Residues, Charged Residues, and Residues with π-Electron Clouds

Here I present an account of utility of the same analysis from amino acid centric (bottom-up) framework.

Considering the centroids (instead of C^α–atoms) of predominantly hydrophobic amino acids as true indicators of positions, magnitude of CD (found from the slope of the log–log plot) between hydrophobic amino acids (Fig. 2.3) within the protein 'Carbon monoxide dehydrogenase' (PDB id: 1su7) was calculated. (As in the case of previous calculation, set of formulae mentioned earlier was used. One can refer to (Banerji and Ghosh 2011) for details of the methodology.) Similarly, considering 'His', 'Arg', 'Lys', 'Asp' and Glu' as the charged amino acids, the correlation dimension between charged amino acids was calculated for the protein 'cold-active citrate synthase' (PDB id: 1a59) (Fig. 2.4). To quantify the self-similarity prevalent in the dipole moment distribution of a complex multi-domain protein 'archaeosine tRNA-guanine transglycosylase' (PDB id: 1iq8), CD was calculated (shown in Fig. 2.5). Equally interesting was the study of CD between π-electron clouds of 'Tyr', 'Trp' and 'Phe', in 'bacterial glucoamylase' (PDB id: 1lf6) (Fig. 2.6). Although Figs. 2.3, 2.4, 2.5, and 2.6 describe the results obtained from individual protein structures, it is easy to note that comprehensive scale-up of these on statistical scale can unearth hidden dependencies between pertinent entities embodying biophysical or biochemical properties.

2.5 Problems with FD, Necessary Precautions While Working with FD

Effective as they are, FD-based measures are not panaceas. There are some inherent limitations associated with FD-based toolkits and whoever interested to explore the scopes of utilization of FD-based measures should be aware of them. For example, as a consequence scale invariance of fractals, the functions describing them can never be smooth. This is significant to note because it implies that the normal practice of expanding a function to first order and then to approach the linearized problem, will not be compatible with FD-based constructs. The scale invariance of fractals ensures that one cannot find a scale where functions become smooth enough to linearlize. Due to this innate nonanalytic nature, fractals have undefined derivatives; making the common principles of differential calculus absolutely useless for them. Thus, the very property that makes fractals useful while describing irregular natural shapes (and phenomena), pose these problems.

Kept below is an account of other specific points that one should be careful about:

1. There are nontrivial differences between proteins and "classical collapsed polymers" (interested reader can refer to page 41–44 of (Dewey 1997)). Hence, some of the theoretical constructs of polymer physics cannot be transported "directly" to the realm of protein structure studies. This is important to note because many of the early works on fractals were carried on polymers and not on proteins. Similarly, while it is known that energy surfaces of proteins are rough hypersurfaces in high dimensional configuration spaces with an immense number of local minima (Frauenfelder et al. 1988); and that proteins share this feature with other complex systems like spin glasses, glass-forming liquids, macromolecular melts etc. (Gallivan and Dougherty 1999); one should not forget that compared to the aforementioned complex systems, a protein is a relatively small system. Thus, a blind application of FD-based procedures pertinent for spin-glass studies on proteins, may not be judicious.

2. One should always be aware of the statistical nature of FD-based measures. This implies that FD-based measures can be useful for studying a property from a "top-down" perspective. Employing the FD-based measures to extract patterns from a "bottom-up" standpoint, where the atoms are distinguishable, can only be done with extreme care, ensuring that the number of distinguishable atoms form a statistically significant set. Thus, one needs to be careful while studying the "local" profile of certain properties with FD measures. Unfortunately, ascertaining the limit of this 'local'-ness in terms of the precise number of (interacting) parameters is primarily dependent upon experience. But a reasonable approach can be to consider a statistically significant number of such interactions. Defining any of the fractal exponents with 5–10 atoms makes no sense.

3. Continuing from the last point, problems with aspects of an interesting parameter averaging out due to the innate constraints of FD-based methodologies, isn't uncommon [see for example (Lee, 2008)] either. Drawing from that, one

can assert that probing protein interior with fractal exponents for some 'nano' or 'microscopic' properties (or emergence of it) may turn out to be extremely risky. Similarly, one should be aware of the possibility that in some cases, a system may show statistical self-similarity only when the X and Y-axes describing it, are magnified by different scales. These fractals form a distinct class of fractal objects and are called the 'self-affine fractals'. While the 'self-affine fractals' are not exactly common, they aren't rare in the paradigm of protein studies either (Glockle and Nonnenmacher 1995). Working with them calls for extra precautions.

4. FD-based analyses often provide results in (somewhat) abstract form that are difficult to be explained fully from our present state of knowledge of biophysics and biochemistry. [Examples of such results can be found in many recent and old works (Cserzoa and Vicsek 1991; Isvoran 2004; Morita and Takano 2009)]. Thus, one should be patient to decipher (and interpret) the abstract nature of fractal results.

2.6 New Directions in Fractal Dimension-Based Protein Interior Research

We have discussed various approaches to calculate fractal dimension for proteins. Results obtained from these studies complement each other. Even a cursory look at the various research-paper repositories will demonstrate the growth in the number of works that are attempting to explore protein interior with fractal exponents. Such growth can be observed to be phenomenal during the last 5–7 years. This body of research, collectively, presents the contour of a promising multifaceted technique that can honestly and reliably probe an object as complex as protein. Indeed a lot has been achieved with it; yet, one can still observe the possibilities of many new works that can still be performed to take the obtained knowledge to completeness. Here, a few simple possibilities are enlisted.

Even after the (pioneering) Leitner work (discussed earlier) (Leitner 2008), one needs to relate the complete information of energy transport channels within individual proteins before relating them to their functional roles. Similarly, studies need to be performed with "universal equation of state" (discussed earlier) (de Leeuw et al. 2009; Reuveni et al. 2008) to fully understand the biophysical and biochemical implications of protein's existence at "the edge of unfolding", before attempting to use it for practical protein engineering applications. The (Reuveni et al. 2008; de Leeuw et al. 2009) paradigm of works can be used to probe more fundamental questions too. For example, it has previously been suggested that, though the theoretical limit on the size of protein structure space is enormously huge (an assertion that indirectly implies that only an infinitesimally small portion of protein structure space has been explored during the course of evolution hitherto), it is quite plausible that most (if not all) of functionally relevant protein sequence space has already been explored (Dryden et al. 2008). Now, put this (extraordinary) claim to present perspective of looking at protein structures. Numerous works of last decade have established the fact that enzymes that are homologous at the sequence level are likely to have the same (or at least a very similar) mechanism. That being the case, employing the (Reuveni et al. 2008; de Leeuw et al. 2009) framework of logic, can one identify and compare the characteristics of aforementioned "universal equation of state" when it describes the functional enzymes and protein structures that are theoretically possible but are not known for any functionality? In other words, can one segregate the functional enzymes from the theoretically possible but functionally silent enzymes based on their proximity to "the edge of unfolding"?—These questions have huge ramifications not only in the sphere of theoretical biology but also in the realm of protein engineering and possibly in the ever growing paradigm of drug-discovery.

Likewise, in another recent work (Banerji and Ghosh 2009b) (discussed earlier), certain nontrivial correlations between polarizability-FD and hydrophobicity-FD in all-α, α/β and (especially) $\alpha + \beta$ proteins, were reported. It may be worthwhile to study whether these correlations imply a small yet significant dependence of dielectric constant on hydrophobicity content of a protein (and vice versa). Similarly one needs to explore the uniqueness of distributions of structural properties in all-β proteins to answer why neither their mass, nor hydrophobicity distribution show any dependence on their polarizability distributions. In this context, perhaps an even more challenging problem will be to ascertain exactly how many atoms are needed to observe the emergence of mass-FD or hydrophobicity-FD or polarizability-FD? Since, as a recent work indicates, emergence of protein's biochemical and biophysical properties with statistical nature, might well be mesoscopic in their nature (Banerji and Ghosh 2010); and since, FD in the context of protein structure studies, essentially quantifies the statistical self-similarity (Mitra and Rani 1993);—evaluation of exact number of atoms to observe the emergence of interior FD parameters will be immensely beneficial.

Questions and points discussed in the paragraph above are far from being simple. However, surprisingly, many of the unsolved problems in the realm of FD-based protein interior studies can be identified with more "simple" questions. For example, as mentioned in earlier discussions, FD of protein backbone is independent of secondary structural content of the protein, which implies that secondary structural classes can reliably be modeled as Hamiltonian walks (Dewey 1997). (Hamiltonian walk is a lattice walk in which every point is visited once and only once, with no intersecting paths. Thus, the construct 'Hamiltonian walk' can be used to model polymers that show compact local and global scaling, viz. "crumpled globules".) Crumpled globules tend to imply an absence of knots in the protein structures (Dewey 1997). But knots do occur in proteins, and their occurrences are not entirely rare (Grosberg and Khokhlov 1994; Røgen and Fain 2003). How does one resolve this contradiction from an objective general perspective?

Talking about an even simpler (?) case, we concentrate on amino acid frequency studies. It was known from an old study that the distribution of pairs of amino acids is not a linear combination of frequencies of occurrences of the constituent individual amino acids (Ramnarayan et al. 2008). A subsequent work could establish that the distribution of a given residue along a polypeptide chain is fractal in nature (Isvoran 2004). However, the same study reported a negative correlation between the probabilities of occurrence of various residues and their fractal dimensions; for which, according to my knowledge, no general satisfactory answer has yet been proposed. Similarly, although a recent study (Morita and Takano 2009) establishes the fact that amino acid residue network in protein native structure forms a fractal and not small-world network, no general biophysical cause to this could be attributed to such observation. But at the same time, their explanation on the reasons behind critically percolated nature of amino acid residue-contact network, could be related to the another assertion made in a recent work (de Leeuw et al. 2009) that says; majority of proteins in the PDB exist in a marginally stable thermodynamic state, namely a state that is close to the edge of unfolding. A comprehensive study involving these two works, viz [(Morita and Takano 2009) and (de Leeuw et al. 2009)]—from the perspective of molecular evolution may provide us with unique perception of protein structures.

2.7   Conclusion

Hence, while a lot of unknown, unexpected and deep information about (nonlinear) dependencies between protein-structural-parameters could be unearthed with FD-based probing, a lot is yet to be achieved. One should never forget that apart from the prominent importance for biological functions, proteins are qualified objects to study the dynamics of complex systems. The FD-based investigations of protein interior organization can serve as a unique platform for the physicists and biologists to interact. However, to end on a cautionary note, although FD-based investigations of protein properties are insightful and although they can often reveal unexpected information, FD-based measures are not panacea. Therefore, one needs to be aware of the intrinsic limitations of them and at the same time take necessary precautions (as outlined before) while implementing them.

References

Abrams CF, Lee NK, Obukhov SP (2002) Collapse dynamics of a polymer chain: theory and simulation. Europhys Lett 59(3):391–397
Agashe VR, Shastry MCR, Udgaonkar JB (1995) Initial hydrophobic collapse in the folding of barstar. Nature (London) 377: 754
Akiyama S et al. (2002) Conformational landscape of cytochrome c folding studied by microsecond-resolved small-angle x-ray scattering. Proc Natl Acad Sci USA 99(3):1329–1334
Alexander S, Orbach R (1982) Density of states of fractals: `fractons'. J Phys (France) Lett 43: L625–L631
Allen JP, Colvin JT, Stinson DG, Flynn CP, Stapleton HJ (1982) Protein conformation from electron spin relaxation data. Biophys J 38:299–310
Amadei A, Linssen ABM, Berendsen HJC (1993) Essential dynamics of proteins. Proteins: Struct Func Gen 17: 412–425

Aqvist J, Luecke H, Quiocho F, Warshel A (1991) Dipoles localized at helix termini of proteins stabilize charges. Proc Natl Acad Sci USA 88:2026–2030

Argyrakis P, Kopelman R (1990) Nearest-neighbor distance distribution and self-ordering in diffusion-controlled reactions. Phys Rev A 41:2114–2126

Aszodi A, Taylor WR (1993) Connection topology of proteins. Bioinformatics 9:523–529

Atilgan AR, Durell SR, Jernigan RL, Demirel MC, Keskin O, Bahar I (2001) Anisotropy of fluctuation dynamics of proteins with an elastic network model. Biophys J 80: 505–515

Bagchi B, Fleming GR (1990) Dynamics of activationless reactions in solution. J Phys Chem 94:9–20

Bahar I, Rader AJ (2005) Coarse-grained normal mode analysis in structural biology. Curr Opin Struct Biol 15:586–592

Bahar I, Atilgan AR, Erman B (1997) Direct evaluation of thermal fluctuations in proteins using a single-parameter harmonic potential. Fold Des 2:173–181

Bahar I, Atilgan AR, Demirel MC, Erman B (1998) Vibrational dynamics of folded proteins: significance of slow and fast motions in relation to function and stability. Phys Rev Lett 80: 2733–2736

Baldwin RL (1989) How does protein folding get started? Trends Biochem Sci 14:291–294

Baldwin RL, Rose GD (1999) Is protein folding hierarchic? I. Local structure and peptide folding. Trends Biochem Sci 24:26–33

Banavar JR, Maritan A (2003) Geometrical approach to protein folding: a tube picture. Rev Mod Phys 75: 23–34

Banerji A, Ghosh I (2009a) A new computational model to study mass inhomogeneity and hydrophobicity inhomogeneity in proteins. Eur Biophys J 38:577–587

Banerji A, Ghosh I (2009b) Revisiting the myths of protein interior: studying proteins with mass-fractal hydrophobicity-fractal and polarizability-fractal dimensions. PLoS ONE 4(10):e7361. doi:10.1371/journal.pone.0007361

Banerji A, Ghosh I (2010) Mathematical criteria to observe mesoscopic emergence of protein biochemical properties. J Math Chem 49(3):643–665. doi:10.1007/s10910-010-9760-9

Banerji A, Ghosh I (2011) Fractal symmetry of protein interior: what have we learned? Cell Mol Life Sci 68(16):2711–2737

Bashford D, Karplus M (1990) pKas of ionization groups in proteins: atomic detail from a continuum electrostatic model. Biochemistry 29:10219–10225

Ben-Avraham D (1993) Vibrational normal-mode spectrum of globular-proteins. Phys. Rev. B 47:14559–14560

Benney B, Dowrick N, Fisher A, Newmann M (1992) The Theory of critical phenomena. Oxford University Press, Oxford

Berman H, Henrick K, Nakamura H (2003) Announcing the worldwide. Protein Data Bank Nat Struct Biol 10:980

Berry H (2002) Monte Carlo simulations of enzyme reactions in two dimensions: fractal kinetics and spatial segregation. Biophys J 83:1891–1901

Bierzynski A et al (1982) A salt bridge stabilizes the helix formed by isolated C-peptide of RNase A. Proc Natl Acad Sci USA 79:2470–2474

Binder K (1983) In phase transitions and critical phenomena, vol 10, Domb C, Lebowitz J (ed). Academic, New York

Blanco FJ, Jimenez MA, Herrantz J, Rico M, Santoro J, Nieto JL (1993) NMR evidence of a short linear peptide that folds into a ß-hairpin in aqueous solution. J Am Chem Soc 115:5887–5888

Blanco FJ, Rivas G, Serrano L (1994) A short linear peptide that folds into a native stable beta-hairpin in aqueous solution. Nat Struc Biol 1:584–590

Bohm G (1991) Protein folding and deterministic chaos: limits of protein folding simulations and calculations. Chaos Solitons Fractals 1:375–382

Bohr HG, Wolynes PG (1992) Initial events of protein folding from an information-processing viewpoint. Phys Rev A 46:5242–5248

Bouchaud J-P, Georges A (1990) Anomalous diffusion in disordered media. Phys Rep 195(4, 5):127–293

Brooks CL, Karplus M, Pettitt BM (1988) Proteins: a theoretical perspective of dynamics, structure and thermodynamics. Wiley, New York

Brown JE, Klee WA (1971) Helix-coil transition of the isolated amino terminus of ribonuclease. Biochemistry 10:470–476

Bryngelson J (1994) When is a potential accurate enough for structure prediction? Theory and application to a random heteropolymer model of protein folding. J Chem Phys 1994(100):6038–6045

Bryngelson JD, Wolynes PG (1987) Spin glasses and the statistical mechanics of protein folding. Proc Natl Acad Sci USA 84:7524–7528

Bryngelson JD, Wolynes PG (1989) Intermediates and barrier crossing in the random energy model (with applications to protein folding). J Phys Chem 93:6902–6915

Bryngelson JD, Onuchic JN, Socci ND, Wolynes PG (1995) Funnels, pathways, and the energy landscape of protein folding: a synthesis. Proteins 21(3):167–195

Buguin A, Brochart-Wyart F, de Gennes PG (1996) Collapse of a flexible coil in a poor solvent. C R Acad Sci Paris Ser II 322: 741–746

Burioni R, Cassi D, Cecconi F, Vulpiani A (2004) Topological thermal instability and length of proteins. Proteins Struct Funct Bioinf 55(3):529–535

Bussemaker HJ, Thirumali D, Bhattacherjee JK (1997) Thermodynamic stability of folded proteins against mutations. Phys Rev Lett 79:3530–3533

Byrne A, Kiernan P, Green D, Dawson K (1995) Kinetics of homopolymer collapse. J Chem Phys 102:573

Bytautas L, Klein DJ, Randic M, Pisanski T (2000) Foldedness in linear polymers: a difference between graphical and Euclidean distances, DIMACS. Ser Discr Math Theor Comput Sci 51:39–61

Carlini P, Bizzarri AR, Cannistraro S (2002) Temporal fluctuations in the potential energy of proteins: noise and diffusion. Phys D 165:242–250

Chahine J, Nymeyer H, Leite VBP, Socci ND, Onuchic JN (2002) Specific and nonspecific collapse in protein folding funnels. Phys Rev Lett 88: 168101

Chan HS, Dill KA (1991) Polymer principles in protein structure and stability. Annu Rev Biophys Biophys Chem 20:447

Chu B, Ying QC, Grosberg AYu (1995) Two-stage kinetics of single chain. Collapse. Polystyrene in Cyclohexane. Macromolecules 28:180–189

Churilla AM, Gottschalke WA, Liebovitch LS, Selector LY, Todorov AT, Yeandle S (1995) Membrane potential fluctuations of human T-lymphocytes have fractal characteristics of fractional Brownian motion. Ann Biomed Eng 24:99–108

Clarkson MW, Lee AL (2004) Long-range dynamic effects of point mutations propagate through side chains in the serine protease inhibitor eglin c. Biochemistry 43:12448–12458

Colvin JT, Stapleton HJ (1985) Fractal and spectral dimensions of biopolymer chains: solvent studies of electron spin relaxation rates in myoglobin azide. J Chem Phys 82:10

Coveney PV, Fowler PW (2005) Modelling biological complexity: a physical scientist's perspective. J R Soc Interf 2:267–280

Creighton TE (1993) Proteins: structures and molecular principles. W. H. Freeman and Co., New York

Cserzo AM, Vicsek T (1991) Self-affine fractal analysis of protein structures. Chaos Solitons Fractals 1:431–438

Cui Q (2006) Normal mode analysis: theory and applications to biological and chemical systems. Chapman and Hall/CRC, FL

d'Auriac JCA, Rammal R (1984) 'True' self-avoiding walk on fractals. Phys A Math Gen 17:L15–L20

Damaschun G, Damaschun H, Gast K, Zirwer D (1999) Proteins can adopt totally different folded conformations. J Mol Biol 291:715–725

Dawson KA, Timoshenko EG, Kuznetsov YA (1997) Kinetics of conformational transitions of a single polymer chain. Physica A 236(1–2): 58–74

de Gennes PG (1979) Scaling Concepts in Polymer Physics (Ithaca. Cornell University Press), NY de Gennes PG (1985) Kinetics of collapse for a flexible coil. J Phys (Paris) Lett 46: L-639–L-642

De Leeuw M, Reuveni S, Klafter J, Granek R (2009) Coexistence of flexibility and stability of proteins: an equation of state. PLoS ONE 4(10):e7296

Delarue M, Sanejouand YH (2002) Simplified normal mode analysis of conformational transitions in DNA-dependent polymerases: the elastic network model. J Mol Biol 320:1011–1024

Dewey TG (1993) Protein structure and polymer collapse. J Chem Phys 98:2250–2257

Dewey TG (1995) Fractal dimensions of proteins: what are we learning? Het Chem Rev 2:91–101

Dewey TG (1997) Fractals in molecular biophysics. Oxford University Press, New York

Dewey TG, Bann JG (1992) Protein dynamics and noise. Biophys J 63:594–598

Dewey TG, Spencer DB (1991) Are protein dynamics fractal? Commun Mol Cell Biophys 7:155–171

Dill KA (1990) Dominant forces in protein folding. Biochemistry 29:7133–7155

Dill KA (1999) Polymer principles and protein folding. Protein Sci 8: 1166–1180

Dill KA, Fiebig KM, Chan HS (1993) Cooperativity in protein-folding kinetics. Proc Natl Acad Sci USA 90:1942–1946

Dill KA, Bromberg S, Yue K, Fiebig KM, Yee DP, Thomas PD, Chan HS (1995) Principles of protein folding–a perspective from simple exact models. Protein Sci 4(4):561–602

Dobson CM (1992) Unfolded proteins, compact states and molten globules. Curr Opin Struct Biol 2:6–12

Doi M, Edwards SF (1988) The theory of polymer dynamics. Clarendon Press, Oxford

Domb C (1969) Self-avoiding walks on lattices. Adv Chem Phys 15:229–259

Doruker P, Atilgan AR, Bahar I (2000) Dynamics of proteins predicted by molecular dynamics simulations and analytical approaches: application to a-amylase inhibitor. Proteins 40:512–524

Drews AR, Thayer BD, Stapleton HJ, Wagner GC, Giugliarelli G, Cannistraro S (1990) Electron spin relaxation measurements on the blue-copper protein plastocyanin: deviations from a power law temperature dependence. Biophys J 57(1):157–162

Dryden DTF, Thomson AR, White JH (2008) How much of protein sequence space has been explored by life on Earth? J R Soc Interface 5:953–995

Duplantier B, Saleur H (1987) Exact tricritical exponents for polymers at the theta point in two dimensions. Phys Rev Lett 59:539–542

Dyson HJ, Rance M, Houghten RA, Wright PE, Lerner RA (1988) Folding of immunogenic peptide fragments of proteins in water solution. II. The nascent helix. J Mol Biol 201:201–217

Eaton WA, Muñoz V, Thompson PA, Henry ER, Hofrichter J (1998) Kinetics and dynamics of loops, α-helices, β-hairpins and fast-folding proteins. Accounts Chem Res 31:745–753

Eichinger BE (1972) Elasticity theory. I. Distribution functions for perfect phantom networks. Macromolecules 5:496–505

Eisenriegler E (1993) Polymers near surfaces: conformation properties and relation to critical phenomena. World Scientific, Singapore

Elber R (1989) Fractal analysis of protein. In: Avnir D (ed) The fractal approach to heterogeneous chemistry. Wiley, New York

Elber R, Karplus M (1986) Low-frequency modes in proteins: use of the effective-medium approximation to interpret the fractal dimension observed in electron-spin relaxation measurements. Phys Rev Lett 56:394

Elber R, Karplus M (1987) Multiple conformational states of proteins: a molecular dynamics analysis of myoglobin. Science 235:318

Ellis RJ, Hartl FU (1999) Principles of protein folding in the cellular environment. Curr Opin Struct Biol 9:102–110

Enright MB, Leitner DM (2005) Mass fractal dimension and the compactness of proteins. Phys Rev E 71:011912

Epand RM, Scheraga HA (1968) The influence of long-range interactions on the structure of myoglobin. Biochemistry 7:2864–2872

Evitt M, Sanders C, Stern PS (1985) Protein normal-mode dynamics; trypsin inhibitor, crambin, ribonuclease, and lysozyme. J Mol Biol 181: 423–447

Family F (1982) Direct renormalization group study of loops in polymer. Phys Lett 92A:341–344

Fan ZZ, Hwang JK, Warshel A (1999) Using simplified protein representation as a reference potential for all-atom calculations of folding free energy. Theor Chem Acc 103:77–80

Feig M, Onufriev A, Lee M, Im W, Case E, Brooks C (2004) Performance comparison of generalized Born and Poisson methods in the calculation of electrostatic solvation energies for protein structures. J Comput Chem 25:265–284

Fermi E, Pasta J, Ulam S (1955) Studies of nonlinear problems. Los Alamos Rep LA-1940, Pap 266: 491–501

Fersht AR (2000) Transition-state structure as a unifying basis in protein-folding mechanism-scontact order, chain topology, stability, and the extended nucleus mechanism Proc. Natl Acad Sci USA 97:1525–1529

Figueiredo PH, Moret MA, Nogueira E Jr, Coutinho S (2008) Dihedral-angle Gaussian distribution driving protein folding. Phys A 387:2019–2024

Finkelstein AV (1991) Rate of β-structure formation in polypeptides. Proteins 9:23–27

Finkelstein AV, Ptitsyn OB (2002) Protein Physics. Academic Press, San Diego

Flory P (1971) Principles of Polymer Chemistry (Ithaca. Cornell University Press), New York

Frauenfelder H, McMahon B (1998) Dynamics and functions of proteins: the search of general concepts. Proc Natl Acad Sci USA 95:4795–4797

Frauenfelder H, Petsko GA, Tsernoglou D (1979) Temperature dependent X-ray diffraction as a probe as of protein structural dynamics. Nature 280:558–563

Frauenfelder H, Parak F, Young RD (1988) Conformational substates in proteins. Ann Rev Biophys Biophys Chem 17: 451–479

Freed KF (1987) Renormalization group theory of macromolecules. Wiley, New York

French AS, Stockbridge LL (1988) Fractal and Markov behavior in ion channel kinetics. Can J Physiol Pharm 66:967–970

Fuentes EJ, Gilmore SA, Mauldin RV, Lee AL (2006) Evaluation of energetic and dynamics coupling networks in a PDZ domain protein. J Mol Biol 364:337–351

Fujisaki H, Straub JE (2005) Vibrational energy relaxation in proteins. Proc Natl Acad Sci USA 102:7626–7631

Fukada K, Maeda H (1990) Correlation between rate or chain folding and stability of the β-structure of a polypeptide. J Phys Chem 94:3843–3847

Gallivan JP, Dougherty DA (1999) Cation–π interactions in structural biology. Proc Natl Acad Sci USA 96:9459–9464

Garcia AE, Blumenfeld R, Hummer G, Krumhansl JA (1997) Multi-basin dynamics of a protein in a crystal environment. Physica D 107:225–239

Gilmanshin R, Williams S, Callender RH, Woodruff WH, Dyer RB (1997) Fast events in protein folding: relaxation dynamics of secondary and tertiary structure in native apomyoglobin. Proc Natl Acad Sci USA 94:3709–3713

Glockle WG, Nonnenmacher TF (1995) A fractional calculus approach to self-similar protein dynamics. Biophys J 68:46–53

Goetze T, Brickmann J (1992) Self similarity of protein surfaces. Biophys J 61:109–118

Go N, Taketomi H (1978) Respective roles of short- and long-range interactions in protein folding. Proc Natl Acad Sci USA 75: 559–563

Go N, Abe H, Mizuno H, Taketomi H (1980) Local structures in the process of protein folding. In: Jaenicke N (ed) Protein folding. Elsevier, Amsterdam, pp 167–181

Goldenfeld N (1992) Lectures on phase transitions and the renormalization group. Addison-Wesley, Reading

Goldstein RA, Luthey-Schulten ZA, Wolynes PG (1992a) Protein tertiary structure recognition using optimized Hamiltonians with local interactions. Proc Natl Acad Sci USA 89:9029–9033

Goldstein RA, Luthey-Schulten ZA, Wolynes PG (1992b) Optimal protein-folding codes from spin-glass theory. Proc Natl Acad Sci USA 89:4918–4922

Goychuk I, Hanggi P (2002) Ion channel gating: a first-passage time analysis of the Kramers type. Proc Natl Acad Sci USA 99:3552–3556

Granek R, Klafter J (2005) Fractons in proteins: Can they lead to anomalously decaying time autocorrelations? Phys Rev Lett 95:098106

Grassberger P, Procaccia I (1983) Measuring the strangeness of strange attractors. Phys D 9:183–208

Grimaa R, Schnell S (2006) A systematic investigation of the rate laws valid in intracellular environments. Biophys Chem 124:1–10

Gromiha MM, Selvaraj S (2001) Comparison between long-range interactions and contact order in determining the folding rate of two-state proteins: application of long-range order to folding rate prediction. J Mol Biol 310(1):27–32

Grosberg AYu, Khokhlov AR (1994) Statistical physics of macromolecules. AIP Press, New York

Grosberg AIu, Shakhnovich EI (1986) (Russian paper) A theory of heteropolymers with frozen random primary structure: properties of the globular state, coil-globule transitions and possible biophysical applications. Biofizika 31(6):1045–1057

Grosberg AY, Nechaev S, Tamm M, Vasilyev O (2006) How long does it take to pull an ideal polymer into a small hole? Phys Rev Lett 96(22):228105

Guerois R, Serrano L (2001) Protein design based on folding models. Curr Opin Struct Biol 11:101–106

Gutin A, Abkevich V, Shakhnovich E (1995) Is burst hydrophobic collapse necessary for protein folding? Biochemistry 34:3066

Gutin AM, Abkevich VI, Shakhnovich EI (1996) Chain length scaling of protein folding time. Phys Rev Lett 77:5433–5436

Hagen S, Eaton W (2000) Two-state expansion and collapse of a polypeptide. J Mol Biol 301:1037

Haliloglu T, Bahar I, Erman B (1997) Gaussian dynamics of folded proteins. Phys Rev Lett 79:3090–3093

Halperin A, Goldbart PM (2000) Early stages of homopolymer collapse. Phys Rev E 61:565–573

Havlin S, Ben-Avraham D (1982a) Fractal dimensionality of polymer chains. J Phys A 15:L311–L316

Havlin S, Ben-Avraham D (1982b) New approach to self-avoiding walks as a critical phenomenon. J Phys A 15:L321–L328

Havlin S, Ben-Avraham D (1982c) Theoretical and numerical study of fractal dimensionality in self-avoiding walks. Phys Rev A 26:1728–1734

Havlin S, Ben-Avraham D (1982d) New method of analysing self-avoiding walks in four dimensions. J Phys A 15:L317–L320

Hayes B (1998) How to avoid yourself. Am Sci 86:314–319

Heath AP, Kavraki LE, Clementi C (2007) From coarse-grain to all-atom: toward multiscale analysis of protein landscapes. Proteins 68:646–661

Helman JS, Coniglio A, Tsallis C (1984) Fractons and the fractal structure of proteins. Phys Rev Lett 53:1195–1197

Henzler-Wildman KA, Lei M, Thai V, Kerns SJ, Karplus M et al (2007) A hierarchy of timescales in protein dynamics is linked to enzyme catalysis. Nature 450(7171):913–916

Herrmann HJ (1986) Comment on fractons and the fractal structure of proteins. Phys Rev Lett 56:2432

Hinsen K (1999) Analysis of domain motion by approximate normal mode calculations. Proteins Struct Funct Genetics 33:417–429

Hong L, Jinzhi L (2009) Scaling law for the radius of gyration of proteins and its dependence on hydrophobicity. J Polym Sci Part B 47:207–214

Ichiye T, Karplus M (1987) Anisotropy and anharmonicity of atomic fluctuations in proteins: analysis of a molecular dynamics simulations. Proteins 2:236–239

Isogai Y, Itoh T (1984) Fractal analysis of tertiary structure of protein molecule. J Phys Soc Japan 53:2162

Isvoran A (2004) Describing some properties of adenylate kinase using fractal concepts. Chaos Solitons Fractals 19:141–145

Jackson SE (1998) How do small single-domain proteins fold? Folding and Design 3:R81–R91

Jacobs DJ, Rader AJ, Kunh LA, Thorpe MF (2001) Protein flexibility prediction using graph theory. Proteins Struct Funct Genet 44: 150–165

Karplus M (2000) Aspects of protein reaction dynamics: deviations from simple behavior. J Phys Chem B 104:11–27

Karplus M, McCammon J (1983) Dynamics of proteins: elements and functions. Ann Rev Biochem 53:263–300

Karplus M, Shakhnovich E (1992) Protein folding: theoretical studies. In: Creighton T (ed) Protein folding. W. H. Freeman, New York, pp 127–195

Kauzmann W (1959) Some factors in the interpretation of protein denaturation. Adv Protein Chem 14:1–63

Kesten H (1963) On the number of self-avoiding walks. J Math Phys 4:960–969

Kim PS, Baldwin RL (1982) Specific intermediates in the folding reactions of small proteins and the mechanism of protein folding. Annu Rev Biochem 51:459–489

Kim PS, Baldwin RL (1990) Intermediates in the folding reactions of small proteins. Annu Rev Biochem 59:631–660

Kim S, Jeong J, Kim YKY, Jung SH, Lee KJ (2005) Fractal stochastic modeling of spiking activity in suprachiasmatic nucleus neurons. J Comp Neurosci 19:39–51

Kitao A, Hayward S, Go N (1998) Energy landscape of a native protein: jumping-among-minima model. Proteins 33:496

Kittel C (2004) Introduction to solid state physics. New York, Wiley

Kitsak M, Havlin S, Paul G, Riccaboni M, Pammolli F, Stanley HE (2007) Betweenness centrality of fractal and nonfractal scale-free model networks and tests on real networks. Phys Rev E 75:056115

Klimov DK, Thirumalai D (1996) Criterion that determines the foldability of proteins. Phys Rev Lett 76(21):4070–4073

Kloczkowski A, Mark JE (1989) Chain dimensions and fluctuations in random elastomeric networks. I. Phantom Gaussian networks in the undeformed state. Macromolecules 22:1423–1432

Klushin LI (1998) Kinetics of a homopolymer collapse: beyond the Rouse-Zimm scaling. J Chem Phys 108(18):7917–7920

Kolb VA, Makeev EV, Spirin AS (1994) Folding of firefly luciferase during translation in a cell-free system. EMBO J 13:3631–3637

Kolinski A, Galazka W, Skolnick J (1996) On the origin of the cooperativity of protein folding: implications from model simulations. Proteins 26(3):271–287

Kondrashov D, Van Wynsberghe A, Bannen R, Cui Q, Phillips G (2007) Protein structural variation in computational models and crystallographic data. Structure 15:169–177

Kopelman R (1986) Rate processes on fractals: theory, simulations, and experiments. J Stat Phys 42:185–200

Kopelman R (1988) Fractal reaction kinetics. Science 241:1620–1626

Korn SJ, Horn R (1988) Statistical discrimination of fractal and Markov models of single channel gating. Biophys J 54:871–877

Kosmidis K, Argyrakis P, Macheras P (2003) Fractal kinetics in drug release from finite fractal matrices. J Chem Phys 119:63–73

Krebs WG, Alexandrov V, Wilson CA, Echols N, Yu H, Gerstein M (2002) Normal mode analysis of macromolecular motions in a database framework: developing mode concentration as a useful classifying statistic. Proteins 48:682–695

Kuriyan J, Petsko GA, Levy RM, Karplus M (1986) Effect of anisotropy and anharmonicity on protein crystallographic refinement. An evaluation by molecular dynamics. J Mol Biol 190:227–254

Lauger P (1988) Internal motions in proteins and gating kinetics of ion channels. Biophys J 53:877–884

Lee CY (2006) Mass fractal dimension of the ribosome and implication of its dynamic characteristics. Phys Rev E 73:042901

Lee CY (2008) Self-similarity of biopolymer backbones in the ribosome. Phys A 387:4871–4880

Lehmann KK, Pate BH, Scoles G (1994) Intramolecular dynamics from eigenstate-resolved intrared spectra. Annu Rev Phys Chem 45:241–274

Leitner DM (2001) Vibrational energy transfer in helices. Phys Rev Lett 87:188102

Leitner DM (2002) Anharmonic decay of vibrational states in helical peptides, coils and one-dimensional glasses. J Phys Chem A 106:10870–10876

Leitner DM (2008) Energy flow in proteins. Annu Rev Phys Chem 59:233–259

Leitner DM, Wolynes PG (1996) Vibrational relaxation and energy localization in polyatomics: effects of high-order resonances on flow rates and the quantum ergodicity transition. J Chem Phys 105:11226–11236

Lewis M, Rees DC (1985) Fractal surfaces of proteins. Science 230:1163–1165

Li HQ, Chen SH, Zhao HM (1990a) Fractal structure and conformational entropy of protein chain. Int J Biol Macromol 12:374–378

Li HQ, Chen SH, Zhao HM (1990b) Fractal mechanisms for the allosteric effects of proteins and enzyme. Biophys J 58:1313–1320

Li MS, Klimov DK, Thirumalai D (2004) Finite size effects on thermal denaturation of globular proteins. Phys Rev Lett 93:268107

Liang J, Dill KA (2001) Are proteins well-packed? Biophys J 81:751–766

Liebovitch LS, Sullivan JM (1987) Fractal analysis of a voltage-dependent potassium channel from cultured mouse hippocampal neurons. Biophys J 52:979–988

Liebovitch LS, Toth TI (1990) Using fractals to understand the opening and closing of ion channels. Ann Biomed Eng 18:177–194

Liebovitch LS, Toth TI (1991) A model of ion channel kinetics using deterministic chaotic rather than stochastic processes. J Theor Biol 148:243–267

Liebovitch LS, Fischbary J, Koniarek JP, Todorova I, Wang M (1987a) Fractal model of ion-channel kinetics. Biochim Biophys Acta 869:173–180

Liebovitch LS, Fischbary J, Koniarek J (1987b) Ion channel kinetics: a model based on fractal scaling rather than multistate markov processes. Math Biosci 84:37–68

Lifshitz IM, Grosberg AY, Khokhlov AR (1978) Diagram of States of the Isotropic Solution of Semiflexible. Macromolecules near the Theta-Point. Rev Mod Phys 50:683

Lockless SW, Ranganathan R (1999) Evolutionarily conserved pathways of energetic connectivity in protein families. Science 286:295–299

Logan DE, Wolynes PG (1990) Quantum localization and energy flow in many-dimensional Fermi resonant systems. J Chem Phys 93:4994–5012

Lois G, Blawzdziewicz J, O'Hern CS (2010) Protein folding on rugged energy landscapes: conformational diffusion on fractal networks. Phys Rev E 81:051907

Lowen SB, Teich MC (1993) Fractal renewal processes. IEEE Trans Info Theory 39:1669–1671

Lowen SB, Liebovitch LS, White JA (1999) Fractal ion-channel behavior generates fractal firing patterns in neuronal models. Phys Rev E 59:5970–5980

Lu M, Ma J (2005) The role of shape in determining molecular motions. Biophys J 89:2395–2401

Lu B, Zhang D, McCammon J (2005) Computation of electrostatic forces between solvated molecules determined by the Poisson–Boltzmann equation using a boundary element method. J Chem Phys 122:214102–214108

Lushnikov SG, Svanidze AV, Sashin IL (2005) Vibrational density of states of hen egg white lysozyme. JETP Letters 82:30–33

Madras N, Whittington SG (2002) Self-averaging in finite random copolymers. J Phys A Math Gen 35:L427–L431

McKenzie DS (1976) Polymers and scaling. Phys Rep (Sect C of Phys Lett) 27(2):35–88

Micheletti C, Lattanzi G, Maritan A (2002) Elastic properties of proteins: insight on the folding processes and evolutionary selection of native structures. J Mol Biol 321:909–921

Millhauser G, Salpeter L, Oswald RE (1988) Diffusion models of ion-channel gating and the origin of the power-law distributions from single-channel recording. Proc Natl Acad Sci USA 85:1503–1507

Mirny L, Shakhnovich E (2001) Protein folding theory: from lattice to all-atom models. Annu Rev Biophys Biomol Struct 30:361–396

Mitra C, Rani M (1993) Protein sequences as random fractals. J Biosci 18:213–220

Miyazawa S, Jernigan RL (1999) An empirical energy potential with a reference state for protein fold and sequence recognition. Proteins Struct Funct Genet 36:357–369

Moret MA, Santana MC, Zebende GF, Pascutti PG (2009) Self-similarity and protein compactness. Phys Rev E 80:041908

Morita H, Takano M (2009) Residue network in protein native structure belongs to the universality class of three dimensional critical percolation cluster. Phys Rev E 79:020901

Moritsugu K, Miyashita O, Kidera A (2000) Vibrational energy transfer in a protein molecule. Phys Rev Lett 85:3970–3973

Moritsugu K, Miyashita O, Kidera A (2003) Temperature dependence of vibrational energy transfer in a protein molecule. J Phys Chem B 107:3309–3317

Munoz V, Eaton WA (1999) A simple model for calculating the kinetics of protein folding from three-dimensional structures. Proc Natl Acad Sci USA 96:11311–11316

Munoz V, Thompson PA, Hofrichter J, Eaton WA (1997) Folding dynamics and mechanism of beta-hairpin formation. Nature 390:196–199

Murzin AG, Brenner SE, Hubbard T, Chothia C (1995) SCOP: a structural classification of proteins database for the investigation of sequences and structures. J Mol Biol 247:536–540

Naidenov A, Nechaev S (2001) Adsorption of a random heteropolymer at a potential well revisited: location of transition point and design of sequencesJ. Phys A Math Gen 34:5625

Nakayama T, Yakubo K, Orbach RL (1994) Dynamical properties of fractal networks: scaling, numerical simulations, and physical realizations. Rev Mod Phys 66:381–443

Neri M, Anselmi C, Cascella M, Maritan A, Carloni P (2005) Coarse-grained model of proteins incorporating atomistic detail of the active site. Phys Rev Lett 95(21):218102

Nicolay S, Sanejouand YH (2006) Functional modes of proteins are among the most robust Phys. Rev Lett 96:078104

Nölting B, Andret K (2000) Mechanism of protein folding. Proteins 41:288–298

Nolting B et al (1997) The folding pathway of a protein at high Resolution from Microseconds to Seconds. Proc Natl Acad Sci USA 94:826

Nonnenmacher TF (1989) Fractal scaling mechanisms in biomembranes. Eur Biophys J 16:375–379

Novikov VU, Kozlov GV (2000) Structure and properties of polymers in terms of the fractal approach. Russ Chem Rev 69:523–549

Ostrovsky B, Bar-Yam Y (1994) Irreversible polymer collapse in 2 and 3 dimensions. Europhys Lett 25:409–414

Pande VS, Grosberg AY, Tanaka T (1995a) Freezing transition of random heteropolymers consisting of an arbitrary set of monomers. Phys Rev E 51: 3381–3402

Pande VS, Grosberg AY, Tanaka T (1995b) How accurate must potentials be for successful modeling of protein folding? J Chem Phys 103: 9482–9491

Pande VS, Grosberg AYu, Tanaka T (1997) Thermodynamics of the coil to frozen globule transition in heteropolymers. J Chem Phys 107:5118

Pande VS, Grosberg AYu, Tanaka T (1998) Heteropolymer freezing and design: towards physical models of protein folding. Rev Mod Phys 72:72–259

Peierls RE (1934) Bemerkungu ¨ber Umwandlungstemperaturen. Helv Phys Acta 7:S81–S83

Petrey D, Honig B (2000) Free energy determinants of tertiary structure and the evaluation of protein models. Protein Sci 9:2181–2191

Pierri CL, Grassi AD, Turi A (2008) Lattices for ab initio protein structure prediction. Protein Struct Funct Bioinf 73:351–361

Plaxco KW, Simons KT, Baker D (1998) Contact order, transition state placement and the refolding rates of single domain proteins. J Mol Biol 277(4):985–994

Poland D, Scheraga HA (1970) Theory of helix-coil transitions in biopolymers. Academic Press, New York

Praprotnik M, Delle Site L, Kremer K (2005) Adaptive resolution molecular-dynamics simulation: changing the degrees of freedom on the fly. J Chem Phys 123: 224106

Privalov PL (1979) Stability of proteins: small globular proteins. Adv Phys Chem 33: 167–244

Ptitsyn OB (1987) Protein folding: hypotheses and experiments. J Protein Chem 6:273–293

Ptitsyn OB, Rashin AA (1975) A model of myoglobin self-organization. Biophys Chem 3:1–20

Qi PX, Sosnick TR, Englander SW (1998) The burst phase in ribonuclease a folding and solvent dependence of the unfolded state. Nat Struct Biol 5:882

Qiu L, Zachariah C, Hagen SJ (2003) Fast chain contraction during protein folding: "foldability" and collapse. Dynamics 90(16):1681031–1681034

Ramakrishnan A, Sadana A (1999) Analysis of analyte-receptor binding kinetics for biosensor applications: an overview of the influence of the fractal dimension on the surface on the binding rate coefficient. Biotechnol Appl Biochem (Pt-1):45–57

Rammal R, Toulouse G (1983) Random walks on fractal structures and percolation clusters. J Phys Lett 44:L13–L22

Ramnarayan K, Bohr H, Jalkanen K (2008) Classification of protein fold classes by knot theory and prediction of folds by neural networks: a combined theoretical and experimental approach. Theor Chim Acta 119:265–274

Rasmussen BF, Stock AM, Ringe D, Petsko GA (1992) Crystalline ribonuclease a loses function below the dynamical transition at 220 K. Nature 357:423–424

Reuveni S (2008) Proteins: unraveling universality in a realm of specificity. PhD thesis. Tel Aviv University, Tel Aviv

Reuveni S, Granek R, Klafter J (2008) Proteins: coexistence of stability and Flexibility. Phys Rev Lett 100:208101

Reuveni S, Granek R, Klafter J (2010) Anomalies in the vibrational dynamics of proteins are a consequence of fractal-like structure. Proc Natl Acad Sci USA 107:13696–13700

Rodriguez M, Pereda E, Gonzalez J, Abdala P, Obeso JA (2003) Neuronal activity in the substantia nigra in the anaesthetized rat has fractal characteristics. Evidence for firing-code patterns in the basal ganglia. Exp Brain Res 151:167–172

Røgen P, Fain B (2003) Automatic classification of protein structure by using Gauss integrals. Proc Natl Acad Sci USA 100:119–124

Rozenfeld HD, Song C, Makse HA (2010) The small world-fractal transition in complex networks through renormalization group. Phys Rev Lett 104:025701

Sangha AK, Keyes T (2009) Proteins fold by subdiffusion of the order parameter. J Phys Chem B 113:15886–15894

Sauder JM, Roder H (1998) Amide protection in an early folding intermediate of cytochrome c. Fold Des 3:293–301

Schellman JA (1955) Stability of hydrogen bonded peptide structures in aqueous solution. Compt Rend Lab Carlsberg Ser Chim 29:230–259

Selvaraj S, Gromiha MM (2003) Role of hydrophobic clusters and long-range contact networks in the folding of (alpha/beta) 8 barrel proteins. Biophys J 84:1919–1925

Sfatos CD, Gutin AM, Shakhnovich EI (1993) Phase diagram of random copolymers. Phys Rev E 48:465–475

Shakhnovich E (2006) Protein folding thermodynamics and dynamics: where physics, chemistry and biology meet. Chem Rev 106(5):1559–1588

Shakhnovich EI, Gutin AM (1991) Influence of point mutations on protein structure: probability of a neutral mutation. J Theor Biol 149:537–546

Sharp K, Skinner JJ (2006) Pump-probe molecular dynamics as a tool for studying protein motion and long range coupling. Proteins 65:347–361

Shi Q, Izvekov S, Voth GA (2006) Mixed atomistic and coarse-grained molecular dynamics: simulation of a membrane-bound ion channel. J Phys Chem B 110:15045–15048

Shlesinger MF (1988) Fractal time in condensed matter. Annu Rev Phys Chem 39:269–290

Shoemaker R et al (1985) Nature of the charged-group effect on the stability of the C-peptide helix. Proc Natl Acad Sci USA 82:2349–2353

Shoemaker K, Kim P, York E, Stewart J, Baldwin R (1987) Tests of the helix dipole model for stabilization of alpha-helices. Nature 326:563–567

Shortle D (1996) The denatured state (the other half of the folding equation) and its role in protein stability. FASEB J 10:27–34

Song C, Havlin S, Makse HA (2006) Origins of fractality in the growth of complex networks. Nat Phys 2:275–281

Sosnick TR, Shtilerman MD, Mayne L, Englander SW (1997) Ultrafast signals in protein folding and the polypeptide contracted state. Proc Natl Acad Sci USA 94:8545

Spassov V, Ladenstein R, Karshikoff AD (1997) A optimization of the electrostatic interactions between ionized groups and peptide dipoles in proteins. Protein Sci 6:1190–1196

Stanley H (1999) Scaling, universality, and renormalization: Three pillars of modern critical phenomena. Rev Mod Phys 71(2): S358–S366

Stapleton HJ (1985) Comment on fractons and the fractal structure of proteins. Phys Rev Lett 54:1734

Stapleton HJ, Allen JP, Flynn CP, Stinson DG, Kurtz SR (1980) Fractal form of proteins. Phys Rev Lett 45:1456–1459

Steinbach PJ, Ansari A, Berendzen J, Braunstein D, Chu K, Cowen BR, Ehrenstein D, Frauenfelder H, Johnson JB, Lamb DC (1991) Ligand binding to heme proteins: connection between dynamics and function. Biochemistry 30:3988–4001

Stepanow S, Chudnovskiy AL (2002) The Green's function approach to adsorption of a random heteropolymer onto surfaces. J Phys A: Math Gen 35:4229

Suel GM, Lockless SW, Wall MA, Ranganathan R (2003) Evolutionarily conserved networks of residues mediate allosteric communication in proteins. Nat Struct Biol 10:59–69

Takens F (1985) On the numerical determination of the dimension of an attractor. In: Braaksma B, Broer H, Takens F (eds) Lecture notes in mathematics, vol 1125. Springer, Berlin, pp 99–106

Tama F, Brooks C (2006) Symmetry, form, and shape: guiding principles for robustness in macromolecular machines Annu. Rev Biophys Biomol Struct 35:115–133

Tama F, Sanejouand YH (2001) Conformational change of proteins arising from normal mode calculations. Protein Eng Des Sel 14:1–6

Tanford C (1961) Physical chemistry of macromolecules. Wiley, New York

Tanford C (1962) Contribution of hydrophobic interactions to the stability of globular confirmation of proteins. J Am Chem Soc 84:4240

Taverna DM, Goldstein RA (2002) Why are proteins marginally stable? Proteins 46(1):105–109

Tejera E, Machadoa A, Rebelo I, Nieto-Villar J (2009) Fractal protein structure revisited: topological kinetic and thermodynamic relationships. Phys A 388:4600–4608

Thirumalai D (1995) From minimal models to real proteins, time scales for protein folding kinetics. J Phys I France 5:1457

Thomas PD, Dill KA (1993) Local and nonlocal interactions in globular proteins and mechanisms of alcohol denaturation. Protein Sci 2:2050–2065

Tirion MM (1996) Low-amplitude elastic motions in proteins from a single-parameter atomic analysis. Phys Rev Lett 77:1905–1908

Tissen J, Fraaije J, Drenth J, Berendsen H (1994) Mesoscopic theories for protein crystal growth. Acta Cryst D 50:569–571

Triebel H (1997) Fractals and spectra, Monographs in Mathematics 91, Birkhôauser

Turner TE, Schnell S, Burrage K (2004) Stochastic approaches for modelling in vivo reactions. Comp Biol Chem 28:165–178

Uzer T (1991) Theories of intramolecular vibrational energy transfer. Phys Rep 199:73–146

Wada A (1976) The alpha-helix as an electric macro-dipole. Adv Biophys 9:1–63

Wagner GC, Colvin JT, Allen JP, Stapleton HJ (1985) Fractal models of protein structure, dynamics, and magnetic relaxation. J Am Chem Soc 107:20

Wang CX, Shi YY, Huang FH (1990) Fractal study of tertiary structure of proteins. Phys Rev A 41:7043–7048

Wang J, Onuchic JN, Wolynes PG (1996) Statistics of kinetic pathways on biased rough energy landscapes with applications to protein folding. Phys Rev Lett 76:4861–4864

Wang J, Plotkin SS, Wolynes PG (1997) Configurational diffusion on a locally connected correlated energy landscape: application to finite, random heteropolymers. J Phys I 7:395–421

Warshel A, Papazyan A (1998) Electrostatic effects in macromolecules: fundamental concepts and practical modeling. Curr Opin Struct Biol 8:211–217

Weisstein EW (2012) Self-avoiding walk. From mathworld–a wolfram web resource. http://mathworld.wolfram.com/Self-AvoidingWalk.html. Accessed 27 Sept 2012

Williams S, Causgrove TP, Gilmanshin R, Fang KS, Callender RH, Woodruff WH, Dyer RB (1996) Fast events in protein folding: helix melting and formation in a small peptide. Biochemistry 35:691–697

Wilson KG (1975) The renormalization group: critical phenomena and the Kondo problem. Rev Mod Phys 47:773

Wilson KG (1979) Problems in physics with many scales of length. Sci Am 241:140–157

Wolynes PG, Onuchic JN, Thirumalai D (1995) Navigating the folding routes. Science 267:1619–1620

Wright PE, Dyson HJ, Lerner RA (1988) Conformation of peptide fragments of proteins in aqueous solution: implications for initiation of protein folding. Biochemistry 27:7167–7175

Xiao Y (1994) Comment on fractal study of tertiary structure of proteins. Phys Rev E 46:6

Yang H, Luo G, Karnchanaphanurach P, Louie T-M, Rech I et al (2003) Protein conformational dynamics probed by single-molecule electron transfer. Science 302:262–266

Yu X, Leitner DM (2003) Anomalous diffusion of vibrational energy in proteins. J Chem Phys 119:12673–12679

Yuste SB, Acedo L, Lindenberg K (2004) Reaction front in an A + B → C reaction-subdiffusion process. Phys Rev E 69:036126

Zana R (1975) On the rate determining step for helix propagation in the helix-coil transition of polypeptides in solution. Biopolymers 14:2425–2428

Zhou Z, Payne P, Vasquez M, Kuhn N, Levitt M (1996) Finite-difference solution of the Poisson–Boltzmann equation: complete elimination of self-energy. J Comput Chem 17:1344–1351

Zwanzig R (1990) Rate processes with dynamical disorder. Acc Chem Res 23:148–152

Zwanzig R, Szabo A, Bagchi B (1992) Levinthal's paradox. Proc Natl Acad Sci USA 89:20–22